The Origin of Life & Humanity

A Layman's View of Creation vs. Natural Phenomena

Joseph Alford,
Biochemical Engineer

ISBN: 978-1-962402-61-3

Library of Congress Control Number: 2022914551

Published by
Fideli Publishing, Inc.
119 W. Morgan St.
Martinsville, IN 46151

www.FideliPublishing.com

All images used herein were obtained from stock.adobe.com unless otherwise noted.

Dedication

This book is dedicated to those who have faith in their beliefs but are open to considering alternatives.

Table of Contents

Preface

The author is a member of the scientific/bioengineering community who spent most of his professional career working with scientists in industry in developing and automating chemical processes utilizing living cells to manufacture pharmaceutical and agricultural products. During this time, he built upon his academic chemical engineering education and progressively learned more about the extraordinary structural and functional complexity of living cells. He has observed, contributed to, and sometimes led, various programs and projects in industry, working with world class scientists, to monitor, analyze, model, modify, control, and improve the ability of living cells to produce desired products.

While progress has sometimes been difficult and slow, with many barriers and challenges (many due to incomplete scientific understanding of cellular operations), sometimes improvements do result, usually attributable to intelligent creative thinking on the part of scientists and engineers. On other occasions, improvements have been achieved from the work of "Mother Nature" via, for example, a scientist submitting to a laboratory an unusual fungus or plant leaf found while hiking in a remote forest during a vacation, resulting in identification of an unfamiliar compound (from the fungus or leaf) that is effective against a pathogen or disease.

The author has frequently heard that the scientific community believes and advocates the natural phenomenon theory regarding the origin-of-life. This theory includes the concept of spontaneous generation, which argues that life arose from non-living matter. This is supported in how the "origin of life" is taught in public schools. This does not represent the belief of many of his scientist and engineer colleagues, nor many philosophers who have authored publications on this subject, nor many members of the American Association for the Advancement of Science (AAAS). So, it seems inappropriate to stereotype the general scientific community as believing a particular "origin of life" theory when that theory is unproven and a large segment of the community believes in something else. The clear conflict between the two major theories (creationism vs. natural phenomena) resulted in the author's interest in exploring the two theories in more depth (but still at a layman's level of understanding); hence this book.

From a religious perspective, tenets of many of the major religions of the world include a God who had much to do with creating life and it is this aspect of creationism (i.e., a creative GOD) that this book will refer to regarding the alternative theory to natural phenomena, that being creationism.

Readers are invited to draw their own conclusion on the origin of life from the preponderance of perspectives, information, data, and quotes contained in this book as well as the many references cited. Caution should exist against putting too much weight on any one observation, calculation, or quote as many involve assumptions. Also, scientific leaders sometimes don't agree with one another on topics they publish on and Bible authors sometimes didn't themselves observe events they wrote about.

The author acknowledges the input to and review of this book by many of his colleagues, including scientists from academia and industry, theologians, engineers, and friends.

Introduction

Objective: To present a brief introduction of a living cell — and also the two main theories regarding the origin of living cells: Natural Phenomena and Creationism.

The Bible, reported as the best-selling book of all time with over 5 billion copies sold (with about 100 million copies sold per year), serves as the moral compass for hundreds of millions of people. It begins in Verse 1:1 of the first chapter (Genesis) stating: "In the beginning, God created the heavens and the earth." Following this verse, the Bible indicates that God created plants, animals, man, and woman. Some people, including some leading scientists, don't think so. So let's examine the relevant evidence and logic.

When considering how life began, there are two main theories. Either life began by natural phenomena; i.e., spontaneous generation from non-living matter, or was created by an intelligent creator — God.

Much has been written (and conjectured) regarding these two theories attempting to explain the origin of life. The topic has been the source of many published books, articles, conferences, and, more recently, computer blogs, especially since Charles Darwin developed his theory of evolution in the mid 1800s.

> **Note #1:** Darwin himself said little about the "origin of life." What subsequently became known as "Neo-Darwinism" is based on Darwin's original theory of evolution that was later adapted to incorporate some of Mendel's theory of genetics and which argues that life began completely from natural chemical reactions (i.e., natural phenomena).

> **Note #2:** In Darwin's day, most cellular components, such as DNA, RNA, mitochondria, and ribosomes, had not yet been discovered and the cell was thought to be a very simple structure. It is now known that nothing could be further from the truth. Prof. W. H. Thorpe (U. of Cambridge), a noted scientist and evolutionist, writes, "The most elementary type of cell constitutes a 'mechanism' unimaginably more complex than any machine yet thought up, let alone constructed, by man."[1]

General Cell Description

Before diving into detail regarding the two main "origin of life" theories, it is instructive to briefly describe the centerpiece of this discussion, the living cell, some examples of which are shown on the following page:

E. COLI BACTERIA   HUMAN CELLS

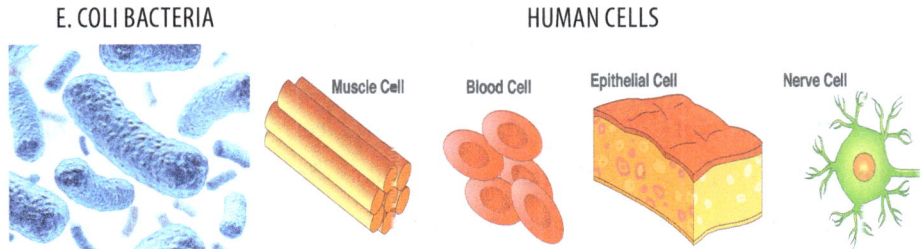

FIGURE 1: EXAMPLES OF CELL TYPES

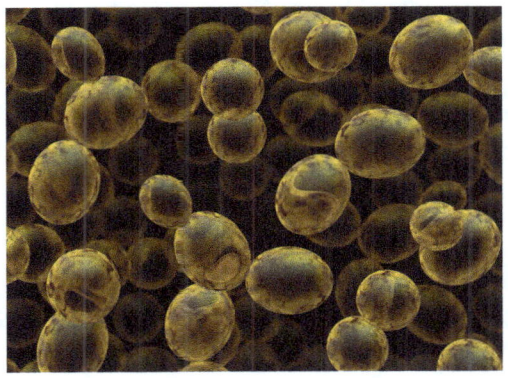

FIGURE 2: YEAST CELLS IN A BIOREACTOR
(fermenting sugars to make alcohol)

Living cells are very small entities. For example, an *E. coli* bacteria cell is about one micrometer (0.00004 inch) in diameter and 2 micrometers in length. Billions of such cells might be contained in each liter of broth in a commercial bioreactor, with an initial small allotment of the cells purposely put into the bioreactor (known as inoculation) which then grow, multiply, and then make a compound destined for a commercial product, such as insulin. *(Note: 1 liter is approx. 1 quart).*

Large concentrations of different microorganisms naturally exist in the air, water, and soil that make up earth's environment. For instance, there are from hundreds to tens of thousands of bacteria, fungus cells, and fungus spores (analogous to seeds of plants) in each cubic meter of the atmosphere.

One scientist's description of a living cell is that it is a high-tech factory, complete with artificial languages and their decoding systems, memory banks for information storage and retrieval, elegant control systems regulating the automated assembly of parts and components, error fail-safe and proof-reading devices utilized for quality control, assembly processes involving the principle of prefabrication and modular construction, and a capability (not equaled in any of man's most advanced machines) of replicating its entire structure within a matter of a few hours.[2]

> **Note:** For *E. coli* bacteria, the replication time is less than an hour.

A cell also controls the shipping and receiving of a great many chemicals, such as nutrients and waste products, through its cell membrane perimeter(s).

This description applies to the large majority of living cells, both single cell microorganisms, such as bacteria, and cells that are part of multicellular organisms, such as mammalian cells in humans. However, there are a few exceptions. E.g., red blood cells do not have a nucleus and so cannot replicate themselves; they are made by bone marrow.

Thus, a cell is extremely complex. A more detailed description of some of the above listed cellular functions is provided later in this book. So, what about the two prevailing theories as to how such cells (i.e., life) came into being?

While evolution is the term most often used when scientists try to describe how humankind came to be, it is generally thought that evolution mostly explains some of the changes that have occurred since life began and does not explain the origin of life itself. Some leading scientists (mostly in academia) argue that natural phenomena was responsible for the origin of life and evolution took over from there, being responsible for certain structural changes to organisms and, presumably (although no explanation exists yet), for such human characteristics as having emotions, consciousness, morality, and spirituality.

Creationists, while acknowledging some role of adaptation in structural changes to organisms (within a species) over time, do not believe natural phenomena explains either the origin of life or, e.g., the spiritual and moral characteristics of humanity, believing instead that both represent evidence of the hand of an intelligent creator (i.e., God).

There are several terms used by non-creationists in discussing the origin of life, including evolution, chemical evolution, Darwinism, Neo-Darwinism, natural events, natural selection, Mother Nature, spontaneous generation, reductionism, and materialism. Instead of dealing with all these terms, this book will usually refer to the term "natural phenomena" as the alternative theory to "creationism."

A common depiction of the two theories is shown in Fig. 3:

GADO CARTOONS

FIGURE 3: THE TWO THEORIES REGARDING EARLY LIFE

FIGURE 4: THE ENVIRONMENT

Lightning strikes onto primordial oceans in a reducing atmosphere contributing to the beginning of life via one of the natural phenomena theories.

Natural Phenomena

Natural Phenomena, sometimes referred to by other terms as mentioned above, argues, in part, that all living species, including humans, evolved from simpler forms of life (i.e., for humans, initially single cells and, eventually, apes) utilizing the natural selection of mutations and inherited variations that increase an organism's ability to compete, survive, and reproduce.

It attempts to theorize, but does not explain how the first living cell(s) came into being- although speculation has suggested that various environmental energy events, such as lightning strikes, occurring over billions of years, allowed simple elements and molecules to eventually combine/react (within a primordial/prebiotic soup — such as in a pond, lake, or ocean), resulting in the sophisticated mix of complex molecules necessary for life. Further, the mix of molecules then somehow organized into highly specific functional modules contained in microscopic aqueous (i.e., water based) volumes surrounded and protected by semi-permeable cell membranes.

Note #3: Natural selection is a process by which species of animals and plants that are best adapted to their environment survive and reproduce, while those that are less well adapted die out. Natural selection is also known as "survival of the fittest."

Creationism

Creationism (ref. the Bible and Quran), sometimes referred to as "theism," argues that an intelligent being (i.e., God) created life and was responsible for creating the first man and woman (Adam and Eve) — including the biological means for them to sustain themselves and reproduce. The Bible does not provide much detail as to how life first began, although it does indicate that plants were created (creation day number 3) before animals (creation days 5 and 6). Also, life in water existed before life on land.

It is unclear if Adam's creation and appearance on earth on creation day number 6 was a single one-time discrete event or the result of sequential events occurring over a period of time. Regardless, the Bible clearly indicates that a creative God directed the appearance of life and humankind. Many verses in the Bible note aspects of God's creativity; some are mentioned later in this book.

Note #4: There is no agreement among scientists or theologians on the actual length of "days" in the 7-day time periods (6 days for creation and 1 for rest) associated with the creation of the world. The Bible (2 Peter 3: 8 and Psalm 90: 4) indicates that a day may have been 1000 years. Separately, the Quran, which is the Muslim's book documenting their history and beliefs, agrees that the earth was created in 6 days via the passage: "Allah created the heavens and the earth, and all that is between them, in six days" (7: 54). The Quran also indicates that a day/time period during the creation of the world is equated to 50,000 years (70: 4) although another verse states that: "A day in the sight of your Lord is like a thousand years of your reckoning." (22: 47).

Note #5: The Quran also describes how "Allah" (i.e., Muslim's God) created Adam (15: 26). The Quran also quotes Allah as proclaiming "We made out of water every living thing (21: 30)."

Note #6: The Muslim's God is not the same as the Christian's God. Christians worship the triune God — i.e., Father, Son, and Holy Spirit. Muslims do not believe in a God with a Son. For Christians, the New Testament of the Bible is clear: To deny the Son is to deny the Father. Such differences in how God is perceived are out of scope of this book. Only God's creative attributes are considered.

Note #7: Many scientists, such as advocates of Neo-Darwinism, view humans the same as other animals in that they evolved from common ancestry (most recently commonality with apes). The Bible teaches that humans are special and resulted from an intentional act of God, making humans in His image. For example, humans alone possess not only body, but also soul and spirit.

Further, advocates of natural phenomena don't believe humans have an individual purpose in life for which a God has played a role. Evolution has no goal in mind (it's a collection of random mutations) and so if the environmental conditions associated with the early earth began again and life actually resulted (which would be unlikely), humans as currently known and structured would undoubtedly not recur in the same way — but some other form of life would result. On the other hand, Christians believe God made people with a purpose in mind and so human's current structure, attributes, and diversity are the direct result of a purposeful design by a Creator.

Creationism is partly based on religious belief, but also gains support from fossils, archaeological records, and the growing knowledge of cell complexity, as well as what believers see as the failures of other theories to explain the evidence.

Most members of many major religions (e.g., Christianity, Judaism, and Islam — known as the great Western religions) believe in creationism as involved in the origin of life. For example, many branches of Christianity regularly recite the Apostles Creed which begins: "I believe in God, the Father Almighty, <u>Creator of Heaven and Earth</u>…".

Also, the Nicene Creed, considered the defining statement of belief of mainstream Christianity (developed in the year 325 and amended in the year 381) begins with: "We believe in one God, the Father the almighty, <u>maker of heaven and earth</u>, of all that is, seen and unseen. …"

On the other hand, leaders and some members of the science and education communities (with many exceptions) are perceived as believing that, somehow, natural phenomena led to the beginning of life. That is, they believe life began by virtue of time, chance, and the inherent (chemical) properties of matter and that no plan, purpose, or creative acts were involved.[3]

Note #8: Just as there are many scientists who don't adhere to the natural phenomena theory of the origin of life, so there are some religions that don't advocate a God as responsible for the "origin of life". For example, Buddhist and Hindu members believe that the world and life have neither beginning nor end. They believe that individual humans pre-exist before their birth on earth and become a new entity after death. This is the doctrine of reincarnation. They believe the average human must go through several lives before achieving spiritual maturity.

Note #9: Agnostics and atheists are among the mix of people with different philosophies of life. They don't believe in a God, but that doesn't necessarily mean they believe in natural phenomena as the origin of life either.

So, different theories exist about the origin of the universe and life, with most people perceived as believing either natural phenomena or an intelligent creator as being responsible — with many people having no opinion or believing in something else.

Unfortunately, while both major theories on the origin of life have large followings, only one is taught in most public schools, which is natural phenomena / evolution. A common form of democracy, which includes "separation of church and state", is one reason for this practice since creationism is considered as driven from the "church." The topic remains controversial as the courts are highly selective in how they interpret "separation of church and state" and many examples exist whereby the US national government (i.e., the federal state) continues to affirm that God the Creator exists and which continues to include God in many of their public activities. The Public Education addendum at the end of this book explores this topic in more detail.

A core issue is that natural phenomena does not satisfactorily explain the origin of life. It is just a hypothesis, unsupported by much relevant data so far, that is being presented in public schools as truth. Nor does natural phenomena explain how humanity acquired emotions, consciousness and morality. Nor does it explain how the universe was created via the Big Bang, but (other than a few summary remarks later in this book) that is a discussion for another day and which is separately discussed in the literature.[4]

This book, in part, uses science-based information and principles, common sense logic, analogies, quotes from well-known scientists and theologians, and a summary of a few published calculated probabilities to discuss the challenges and barriers regarding the only non-creation theory that has appealed to a significant portion of the population in explaining the origin of life, that being natural phenomena. Thus, this book encourages readers to "keep an open mind" and consider an intelligent creator as a viable alternative.

This book does so using concepts that most readers (especially ones who have had high school level biology and chemistry courses) can understand and relate to. The book differs from many pro-creation and pro-evolution articles and books in the literature that go into difficult to understand detail regarding statistics, biological/chemical phenomena and deep interpretations of scripture, using nomenclature, descriptions, and arguments that many readers find hard to understand.

While trying to keep this book's content relatively easy to understand, it is still necessary to spend a portion summarizing the extraordinary complexity of even the simplest cells — since believing in natural phenomena as responsible for the origin of life is to agree that such complexity could have happened without an intelligent creator with a purpose and plan.

The content of this book does not deny or question the existence of natural phenomena / evolution (at least with regards to changes that occur within a species). The contention is that it does not satisfactorily explain the origin of life (or of humanity) nor does it present any reasonable rational probability of it being correct, leaving creationism as the remaining viable alternative theory. And, of course, the Bible, known as God's Word to mankind, which was authored by prophets, apostles, and others inspired via the Holy Spirit, teaches that divine creative guidance was involved in the origin of life.

Given the history of science (in which initial theories are not always correct), and given that in science, a healthy skepticism is a professional necessity, it is something of a mystery as to why many leading scientists are so set in believing only one unproven statistically improbable theory regarding the origin of life and are not more open to other possibilities. For example, why is it that some scientists can't believe that there is any entity that is smarter than man? One wonders how man acquired his awesome ability to intelligently create if a more intelligent creator doesn't exist.

Most readers will readily agree as to the creative ability of humanity — and that complex non-biological systems (e.g., automobiles, airplanes, buildings, computers) that have been devised and constructed since human's arrival on earth have required human's intelligent creative involvement. It is instructive that, as smart as mankind is, no human or scientific team (not even the world's most brilliant scientists) have come close to creating life in the laboratory from non-living matter. Could it be that a God who is an intelligent creator exists, who guided the processes that brought forth life on this planet? If readers agree with or are open to this possibility, they should read on.

CHAPTER 1

Does Science Always Get It Right?

Objective: To show that science, in describing phenomena in nature, does not always initially get it right, especially when data is sparse and proofs do not exist. Might this be the case with the natural phenomena theory regarding the origin of life?

As good as science is (and it is generally very good), it is not infallible. There are several well-known examples where the science world has drawn initial conclusions about a topic based on available information and seemingly reasonable assumptions, with conclusions later modified or corrected as additional evidence was collected. For example, scientists (originally

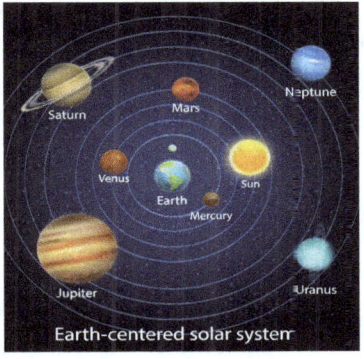

FIGURE 5: EARLY SCIENCE (GREEK) VIEW OF THE UNIVERSE

Greek) once believed and taught a geocentric view of the universe in which Earth was at the center — and the sun, moon, and planets orbited the Earth.

Later on, the famous astronomer Copernicus championed the theory that our Sun was the center of the universe (also later proven untrue when relevant data became available). Even later, as understanding of our solar system improved, orbits of planets around the sun were believed to be circular. Johannes Kepler in the 1600s then showed that the orbits are not circular but elliptical.

As another example, students used to be taught (and sometimes still are) that matter only exists in three states: gas, liquid, and solid. This despite the existence of the 4th state of matter, plasma, which makes up over 99% of the matter in the universe. The existence of plasma was discovered by Sir William Crookes in 1879 with many physicists (e.g., Langmuir) pursuing research and publishing on plasma since the early 1900s.

Plasma is usually created when a gas is heated — which causes it to lose electrons (i.e., ionize) which changes many of its properties. That is, individual particles in a gas are generally neutral, while those in a plasma are electrically charged, making them conductive and responsive to magnetic fields. Examples of plasma include lightning strikes, auroras, welding arcs, light in neon and fluorescent tubes, the earth's ionosphere, solar wind, fire, and the content of most of the stars, including our Sun.

Scientists also once believed that sunlight (with its radiant energy) was needed to support life. Later, living microbes such as bacteria were discovered in hydrothermal vents in ocean bottoms (where sunlight doesn't reach) — and also deep in the earth's crust.

Scientists also once believed that all viruses were either benign or harmful, but now realize many viruses play key beneficial roles in life. That is, scientists once thought of humans as made up of mammalian cells that are occasionally invaded by microbes. Now they view humans as super-organisms of cohabitating mammalian cells, bacteria, fungi and viruses.

Scientists, including the famous Sir Isaac Newton, once believed base metals, such as lead, could be converted into gold. This belief, along with the pursuit of a universal cure of disease and the means of indefinitely prolonging life, were parts of a field known as alchemy. The objectives of alchemy were based on a misunderstanding of basic chemistry and physics, and are now known to be unobtainable.

Also, scientists, up until at least the late 19th century, believed that space was not a vacuum, but filled with an "ether." This was to accommodate their theory at the time that light is a wave that travels through space. That is, since waves cannot travel in empty space, space could therefore not be empty.

Today, scientists understand that light exhibits both wave and particle characteristics and it is the particle characteristics that explains why light does not need a medium in which to travel through. And, until Albert Einstein came along, scientists believed and taught that light always travels in a straight line and that time could not be compressed or expanded. Einstein correctly predicted that light can be bent (such as when traveling close to massive bodies in the universe) and certain situations in the universe can speed up or slow down time.

Speaking of Einstein, one of his famous quotes is: "The More I study Science, the More I Believe in God." Einstein is perceived by some as an atheist; however, he was known to believe in God, the Creator, just not a personal God.[5]

There are other topics in which the opinions of representatives of the science community (e.g., the American Association for the Advancement of Science — AAAS) differ significantly from large segments of US adult opinion. For example, according to the Pew Research Center which has conducted surveys of both groups (with a +/–3% sampling error), over 33% of US adults disagree with AAAS scientists on topics such as human evolution, whether it is safe to eat genetically modified foods, whether it is safe to eat foods grown with pesticides, and whether humans are responsible for climate change.[6]

More recently, during the 2020-2022 Corona virus (Covid-19) pandemic (which officially claimed over 7 million lives world-wide with the actual number thought to be at least 15 million), the credibility of leading science driven health organizations — including the World Health Organization (WHO, which directs and coordinates health efforts within the United Nations) became an issue.

Concerns with the WHO included confusing and conflicting guidance in public statements, delays in declaring a global emergency, delays in recommending travel restrictions (resulting in rapid spread of the virus), and lack of independence from China (the original source of the virus). These concerns ended up hampering responses by countries to the crisis and affecting many people's willingness to take the vaccines that were subsequently offered (for free).

So, science's "getting it right" involves not only obtaining the right data, but also in correct interpretation of the data, clear communication of results, educating the public, and taking appropriate data-driven timely actions. That often happens; but, as with Covid-19, not always.

These examples are only included here to note that the beliefs, recommendations, and teachings of scientists are not always based on valid experimental data and correct interpretation of results. In the case of the "origin of life," it is believed that the views of some scientists are based primarily on their refusal to even consider that a God exists.

> **Note:** It is important to avoid stereotyping scientists as not believing in God. In a published 2009 study by the Pew Research Center, confirmed in a 2018 update, having interviewed the members of the American Associ-

ation for the Advancement of Science (AAAS), their survey indicated that 33% of scientists believe in GOD, and that an additional 18% believe in a "universal spirit or higher power." This contrasts with the general population in which 90% believe in GOD or a higher power.[7]

Regardless, it appears that the scientific community is split roughly 50:50 on whether a God (or higher power) does or does not exist. It is certainly noteworthy that scientists can look at a body of evidence and draw radically different conclusions. Science isn't supposed to work that way. Different theories — yes, but not highly conflicting conclusions.

What does the US National Academy of Sciences say?

The prestigious US National Academy of Sciences has gone on record stating: "Science is a way of knowing about the natural world. It is limited to explaining the natural world through natural causes. Science can say nothing about the supernatural. Whether God exists or not is a question about which science is neutral."[8, 9]

So, if science is officially neutral on the possibility of a divine creator involved with the origin of life (or any other creative activity), why do many leading scientists (and educators) insist on putting only an unproven and unlikely alternative theory into a school's science curriculum and teaching it as fact?

CHAPTER 2

The Big Bang and Conditions of the Universe Necessary for Life

Objective: Review the scientific method and examine the environment that makes life possible.

B efore reviewing the biology and chemistry of living things, it is appropriate to examine the environment that makes life possible. The author believes that scientists, with few exceptions, have generally done an excellent job in evolving their understanding of the universe since the 1600s. In doing so, the standard "scientific method" has usually been followed in proposing theories (i.e., hypotheses) and then conducting appropriate experiments (to prove or refute the hypothesis). The method is shown in Fig. 6.

A classic example in the use of the scientific method was the hypothesis that black holes exist in the universe.

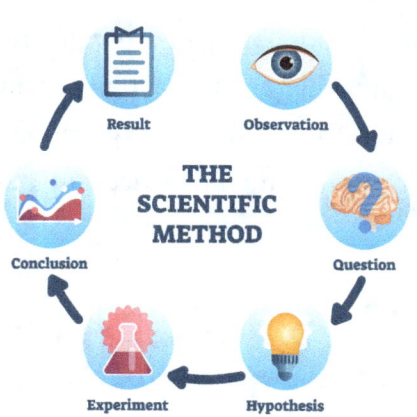

FIGURE 6: THE SCIENTIFIC METHOD

Note: Black holes are thought to be original massive stars that collapsed (that is, became extremely dense) after running out of their nuclear fuel.

This not obvious and non-intuitive hypothesis (to most people) was originally suggested by John Michell in 1783 with the idea remaining hidden until more formally theorized by Karl Schwarzchild in 1916, based on some of Einstein's 1915 published work on relativity. The first label attached to such stars was "frozen stars" with the name later changed to "black holes." Later, other scientists enhanced the theory of black holes (e.g., Oppenheimer, in 1939, also making use of Einstein's previous work with relativity).[10]

World famous British physicist Steven Hawking also added to the understanding of black holes. However, the existence of black holes remained an unproven theory for a long time, and was discussed and taught as such, until their existence was proven (in about 1971) with designed experiments and confirming data.

Another example is the development of the atomic theory. This theory, including the existence and structure of atoms, was hypothesized long before chemical experiments and instruments such as electron microscopes were able to (indirectly) confirm the existence of atoms. *Note: atoms have not been seen directly (such as with optical microscopes) as they are too small; but they have been proven to exist by indirect means.*

Another example regards the Higgs Boson. This subatomic particle, (named after Nobel Prize winning British theoretical physicist Peter Higgs theorized its existence in 1964), subsequently became known as "The God Particle." This was because it existed in theory but was never detected with instrumentation (for over 45 years). So, the existence of the particle was accepted for a long time by the scientific community "on faith" much like the existence of God is accepted by most religious people "on faith." This changed in 2012 when the particle was finally proven to exist by physicists, but the name "The God Particle" has stuck, even becoming a question and answer item on the popular TV show *Jeopardy* in April, 2021.

> **Note:** the Higgs Boson has not actually been detected directly itself, but it is now known to exist briefly as a result of atomic particle collisions (in an accelerator) with it then decaying (disintegrating) almost immediately into fragments which can be detected by instruments.

So, while the Higgs Boson may no longer be a good example of something existing but unproven, it was for a few decades, and the belief that other existing but unseen phenomena exists remains a tenable scientific paradigm.

Returning to "the origin of life" topic, it should still be in the hypothesis stage as regards science. That is, as of 2024, natural phenomena causing the origin of life should be considered just a theory. There is little supporting data or proof as to its being true. So, where are the experimental results that show that living cells can be created via natural means from non-living matter? Confirming experimental results don't exist, so life beginning from natural phenomena is just an unproven theory and not fact.

The Big Bang

The conditions for life ultimately began with the beginning of the universe, often referred to as the Big Bang. The Big Bang is the prevailing cosmological model explaining the existence of the observable universe from the earliest known periods of time through its subsequent large-scale expansion and evolution. The model explains how the universe expanded from an initial state of high density and temperature and offers a comprehensive explanation for a broad range of observed phenomena, including the abundance of light elements, the cosmic microwave background, and large-scale structure.[11]

Steven Hawking, one of the greatest physicists of all time, who was an evolutionist and atheist, has stated (in discussing the Big Bang as the beginning of the universe) : "So long as the universe had a beginning, we could suppose it had a creator."[12] However, he separately indicated he did not believe in a God, claiming instead, via his model of cosmology that the universe did not have a finite beginning (i.e., the Big Bang) — since such a conclusion causes a singularity (that is, irregularity or discontinuity) in his mathematical model of the universe.

To overcome the Big Bang singularity (in which something was seemingly created from nothing), his model needed to use certain "imaginary numbers" rather than real numbers, which can be an issue when trying to describe reality. Regardless, modern scientists now believe that the Big Bang was real and that Hawking's theory of cosmology is not accurate with respect to the beginning of the universe.

Well known Canadian Astrophysicist, Dr. Hugh Ross states: "If the universe arose out of a big bang, it must have had a beginning. If it had a beginning, it must have had a Beginner" (i.e., Creator).[13]

Another atheist who weighed in on the origin of the universe was well-known astronomer Dr. Carl Sagan, Professor at Cornell U. and Director of the Laboratory for Planetary Studies. He also realized the need to avoid the mathematical "singularity" issue represented by the Big Bang and so he proposed that the universe has been infinite in time and space (so did not have a specific beginning) and that it constantly oscillates over time (sometimes expanding and sometimes contracting — like a swinging pendulum).

This theory has subsequently been proven invalid — as the universe is now known to be constantly expanding. Projecting (i.e., extrapolating) the motion of the galaxies back in time (i.e., from knowing the current mapping of celestial bodies and the rate of the expanding universe) results in celestial bodies coming together at a specific point in time, roughly 14 billion years ago, known as the Big Bang.[14] Also, the discovery of cosmic microwave background radiation in the universe, which is direct evidence supporting the Big Bang theory, earned Arno Penzias and Robert Wilson a shared 1978 Nobel Prize in physics.

An interesting aspect of the now accepted "expanding universe" theory is that the first theoretical scientific evidence for an expanding universe (initiated with a "big bang") occurred in 1916 when Albert Einstein noted that his equations of general relativity predicted an expanding universe. Unwilling to accept this conclusion, Einstein altered his theory (adding what became known as the Cosmological Constant) to align his theory with the common wisdom of his day which believed the universe has been eternally existing.[15]

Einstein was later known to have greatly regretted his decision to add the Cosmological Constant into his equations — which suggests that even the greatest scientists sometimes bow to the prevailing consensus of colleagues, including when scientist bias against religion is involved. With the help of data from NASA's COBE satellite many years later, the scientific community then concluded the "big bang" was real and that the universe is continually expanding. So, it seems Einstein's original "theory of relativity" equations were correct; he just didn't accept at the time the portion of his theory's predictions that conflicted with prevailing scientific belief.

Note that the Bible appears to imply this same conclusion (i.e., an expanding universe) when, over 2500 years ago, in Isaiah 42: 5, it is stated, "This is what the Lord says — He who created the heavens <u>and stretched them out</u>." Five different Bible authors (Moses, David, Job, Isaiah, and Jeremiah), in eleven different verses, make a statement regarding the universe "stretching."[16]

So, how did biblical prophets 2500 years ago believe the universe was expanding (i.e., stretching out) when no scientific supporting data existed (or was even possible to determine at the time, given the constraints/interferences of earth's atmosphere)? Could it be that an intelligent creator existed who communicated (perhaps spiritually) revelations to his prophets which were then recorded in what eventually became the Bible?

Islam's Quran also notes: "We expand the sky or the universe to a great extent." This is what science has concluded today and what the Quran implied as reality centuries before the invention of the first telescope.

There are other parts of the Bible that appear to accurately predict aspects of the universe although no scientific support for them existed at the time. For example, Isaiah 40: 22 (written about 700 BC) indicates that the earth is round in the verse "God sits above the circle of the earth."[17] The first confirmation of the Earth's shape didn't occur until ancient Greek philosophers concluded the existence of a spherical Earth in their writings that occurred 200 years later (about 500 BC), with scientific confirmation (by Greek Astronomers) not occurring until about 300 BC.

> **Note:** While many educated people understood the earth was round from BC times, many others (including many seafarers) believed, up until medieval times, that the world was flat. In fact, belief by some that the earth was flat was one of the difficulties Christopher Columbus

encountered in getting approval and funding for his quest to find a route to India and the Far East in the 1490s.

Anyway, the Bible seems to have indicated the correct Earth shape long before the first confirming scientific data was obtained. So how did God's prophet Isaiah know this?

A well-known quote from Sir Fred Hoyle, a famous English astronomer and mathematician, is: "The big bang theory requires a recent origin of the Universe that openly invites the concept of creation."

> **Note #1:** The existence of the Big Bang is a major problem for atheist scientists, as the universe having a specific beginning raises the question of "what caused the Big Bang" (i.e., an event must have a cause). "Natural phenomena" has no causal explanation for the Big Bang.

The above comments on the Big Bang represent yet another example that scientists don't always start out with the correct understanding of the world around them. Such may be the case with their current theories on the origin of life.

A Finely Tuned Universe

The Big Bang was just the first step in creating the universe in its current form. Many properties of the expanding universe in general, and the earth in particular, were then needed for life to exist. A problem for atheists is in explaining how random or natural events explain the super precise fine tuning of the universe and earth that enables and provides conditions for life.

There are several fundamental physical properties of the universe (e.g., gravity, electromagnetism, the energy density of empty space, the expansion speed of the universe, strengths of the basic forces of nature, the material content of the universe, the size and charge of electrons and protons in atoms), many of which if slightly altered would result in no stars, no complex elements, and no life.[18]

In exploring two of these physical properties further, first consider gravity. Without the force of gravity to hold the sun together, the intense pressures at its core would cause it to burst open in a titanic explosion. The same thing would happen to all the other stars in the Universe. Eventually there would be no clumps of matter (e.g., stars or planets) anywhere in the Universe. Without gravity, there would be no atmosphere on earth (or anywhere else) — so there would be no air to breathe.

FIGURE 7: THE EARTH'S MAGNETIC FIELD

Regarding electromagnetism, the Earth's magnetic field serves to deflect most of the solar wind, whose charged particles would otherwise strip away the ozone layer that protects the Earth from harmful ultraviolet radiation. Without the magnetic field, life on Earth would be over very quickly. Auroras (known as northern lights) are the result of disturbances in the magnetic field (i.e., magnetosphere) caused by the solar wind.

The documented list of critical universal constants, required for the universe and life to exist, has grown over time. The list was originally six such constants, then the list grew to 26 and is now over 30.[19] Some of the additions have included the earth's distance from the sun, the sun's size and radiation emissions of the right wavelengths (i.e., colors, UV radiation), the sun's composition (such as the abundance of heavy metals), the sun's stability (with very little variation in light intensity), the earth's size, the role of the moon in stabilizing the earth's axis and its major contribution to the tides that flush and circulate the oceans, and the nearly circular nature of the earth's orbit (vs. a more eccentric elliptical orbit such as that of Mercury) which helps the Earth avoid major extremes in weather.

For example, if the Earth's distance from the sun were changed by 5% either way, animal life (as we know it) would be impossible.[20] Also, if earth was much closer to the sun, the liquid water on earth, including the oceans, would evaporate. Therefore, the extremely precise requirements for a universe allowing for its own existence, as well as for life, argues for the existence of a creator.

There are also other conditions seemingly required for life as humans know it, although not necessarily associated with the above-mentioned universal constants. One example is the amount of carbon dioxide (CO_2) in the atmosphere. CO_2 is a greenhouse gas which means its existence helps reduce the loss of heat from the earth to outer space. Without greenhouse gases, which help the earth retain heat, the temperature of the earth would drop significantly.

NASA has estimated that, without CO_2 in the atmosphere, the average temperature on earth would be about 60 degrees F. cooler. The concentration of CO_2 in the atmosphere, normally about 0.04% [400 parts per million (PPM)] has been increasing by 2 PPM per year in recent decades, with the increase perceived as caused by the burning of fossil fuels like coal, wood, natural gas, and petroleum products, which generates CO_2.

Scientists believe the increase in CO_2 has been a major contributor to global warming (with all its well-known negative consequences), with the average temperature at the earth's surface increasing in recent decades by about 0.33-degree F. per decade. This example, frequently discussed in the national news, illustrates how a small change in a critical atmospheric gas concentration (2 PPM/year in the case of CO_2) can have a significant negative impact on life on earth.

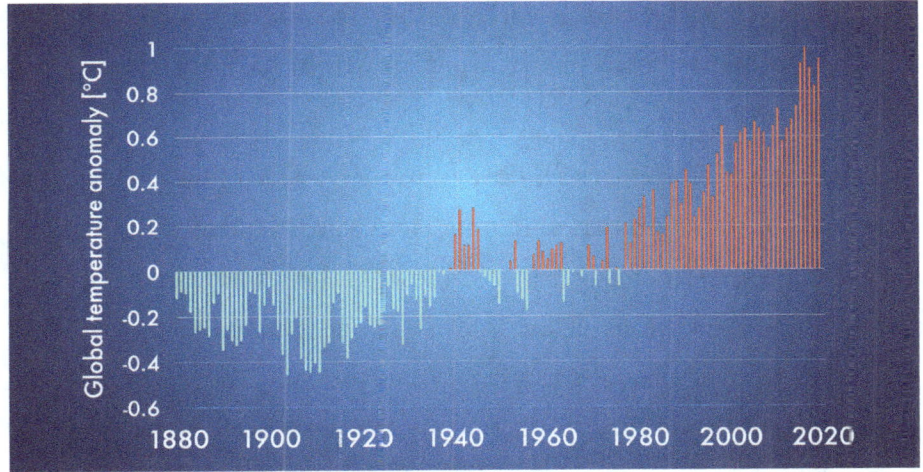

FIGURE 8: GLOBAL WARMING

Another example of a factor affecting the possibility for life on earth is the two Van Allen radiation belts that exist outside much of the earth's atmosphere. These belts are zones of energetic charged particles, most of which originated from the solar wind (i.e., the sun), which have been captured and held by the earth's gravity and magnetic field. They exist from about 40 miles to 36,000 miles above much of the earth's surface (with the exception of the poles). These belts help protect Earth from harmful solar radiation. Without them, there would be significant atmosphere loss.

For example, Mars, with 35% of earth's gravity and no radiation belts has only about 1% as much atmosphere as does the earth. Many celestial bodies, such as Mercury, Venus, Mars, and the Earth's moon, do not have such radiation belts, but the existence of such belts is critical in enabling life, as we know it, on earth.

> **Note #2:** Some planets do have a radiation belt.

> **Note #3:** The ozone layer in the upper atmosphere (much closer to the earth's surface than the radiation belts) also helps protect living organisms from harmful solar radiation.

There does not appear to be any controversy about how critical these universal constant numbers and other factors are to establishing an environment suitable to life. The only argument atheists have come up with is that there are billions of different stars and planets in the universe and odds are that at least one planet (and probably more) had/has the winning formula to support life. Said another way, Earth was apparently incredibly fortunate to have drawn the lucky lottery ticket (or one of a few tickets) in having the correct numerical values and precise combination of many factors (such as physical and chemical conditions within the universe and on earth) to enable life.

More and more scientists are studying the mind-boggling convergence of so many extraordinary "coincidences" in the cosmos that make intelligent life possible on Earth and concluding that this can't possibly be an accident. They're seeing signs of design.[21]

Dr. Paul Davies, noted Professor, Physicist, and Officer in the International Academy of Astronautics, has stated "The origin of life is one of the great outstanding mysteries of science."

Famous rocket scientist, Dr. Wernher von Braun has stated: "I find it as difficult to understand a scientist who does not acknowledge the presence of a superior rationality behind the existence of the universe as it is to comprehend a theologian who would deny the advances of science."[22]

Science leads us to its boundaries where it introduces us to philosophy. It tells us that the laws of nature exist (e.g., gravity), what those laws are, and what they accomplish. But science can't tell us why these laws are as they are — for that we need something else (God, an agent, or ? ?).[23]

CHAPTER 3

Early Earth's Atmosphere and the Miller Experiment

Objective: Examine the credibility of the historical experiment that scientists often point to in support of their natural phenomena theory on the origin of life. Review aspects of the likely composition of earth's early atmosphere and how the assumed composition would affect the natural phenomena theory on how life began.

Many brilliant scientists have spent their careers working on theories and experiments to try and help explain the origin of life, especially since Watson and Crick determined (in the 1950s) the existence, importance, and structure of DNA (deoxyribonucleic acid) as the repository of the genetic code. However, DNA, or even the simpler molecule related to the genetic code, RNA (ribonucleic acid), has yet to be produced by anyone from scratch in the laboratory. Further, the scientific community is still struggling to explain how the first cells could have occurred to simultaneously accomplish three required conditions for life:

1) the need to conduct metabolism (i.e., the ability to capture energy and use it to conduct cellular operations and make cellular material),

2) cell replication/reproduction, and

3) the need to keep all the necessary cell components (i.e., chemicals and subsystems) together inside a container (i.e., semi-permeable cell membrane) in which the "container" allows selective transport of nutrients and waste products between the inside and outside of the cells.

So, advocates claiming natural phenomena as responsible for the origin of life have a long way to go to show that life could reasonably have begun without at least some involvement from an intelligent creator.

The results of selected experiments have encouraged naturalists and evolutionists to keep working on their theories. For example, a scientist (Stanley Miller), with the help of a colleague (Harold Urey) in 1952, mixed four chemicals together that were suspected as being present on the early Earth: boiling water, hydrogen gas, ammonia, and methane.

> **Note #1:** The gas mixture used by Miller-Urey was probably based on published work by Oparin and Haldane in the 1920s that postulated that the early earth's atmosphere contained reducing gases such as hydrogen, ammonia, and methane — but no oxygen).

> **Note #2:** The mixture used by Miller-Urey is "reducing" in nature (vs. oxidizing) — which represents a chemical environment that enables synthesis (i.e., generation) of new and larger molecules such as amino acids- which are the building blocks for proteins.

The gases were put in a container (a glass flask) and subjected to repeated electric sparks to simulate the lightning strikes (i.e., energy source) that would have been a common occurrence on Earth long ago.

By the end of a week, the water in the flask was deep red and turbid. Clearly, a mix of new chemicals had formed. Laboratory analysis showed that the mix contained two amino acids (glycine and alanine), and possibly one or two more of the 22 different amino acids that commonly exist in cells. Their experiment probably produced a few more amino acids than this, but in such tiny quantities that laboratory analysis available at the time could not detect them. Amino acids are often described as the building blocks of life. They are used to form the proteins that control most biochemical processes in living cells. So, Miller and Urey had made at least two of life's many important components from scratch, which was a good beginning. This frequently referenced work was published in1953.[24]

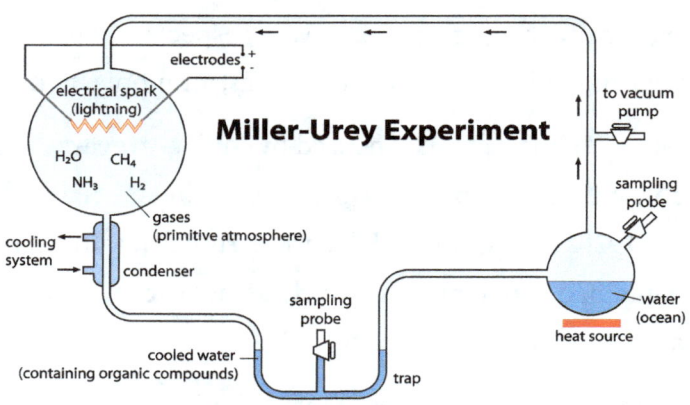

FIGURE 9: GRAPHIC ILLUSTRATING MILLER AND UREY'S EXPERIMENT

However, it was later shown that the mix of gases used by Miller and Urey was not really representative of the mix of gases that existed on early Earth. That is, the early primeval atmosphere was actually hydrogen poor and is now believed by scientists to have been mostly nitrogen and carbon dioxide. Any hydrogen present (due to its light weight) would have quickly escaped into outer space and any ammonia present would have been quickly destroyed by UV light. If significant methane had been present, this would have resulted in early rocks containing significant amounts of carbon or organic compounds, and no geological evidence of this has been found.[25]

> **Note #3:** The mix of gases used in experiments and whether the mixture is oxidizing or reducing is the subject of ongoing conjecture and is important in assessing the probability of natural phenomena creating the building blocks of life. That is because only a reducing atmosphere could lead to the generation of amino acids critical for life. The following section and others in later portions of the book examine this aspect of the conditions necessary for life.

> **Note #4:** The original Miller-Urey gas composition was reducing in nature since hydrogen, methane, and ammonia are reducing agents. However, if the nitrogen-CO_2 composition is more correct, this would represent a slightly oxidizing environment as CO_2 is a mild oxidizing agent and N_2 is generally inert under atmospheric conditions. If some N2 does react with other molecules, it could be either an oxidizing or reducing agent, depending on the specific reactions involved.

> **Note #5:** Based on current estimates of primordial atmospheric composition, it is possible (and perhaps even likely) that a low concentration of oxygen was present in the early atmosphere, contributing (along with CO_2) to a slightly oxidizing environment. Evidence includes sedimentary rock composition data, measurements of

atmospheres of certain celestial bodies, and the known generation of oxygen from the reaction of UV light with water vapor. Some scientists have argued that oxygen did not exist in the early earth's atmosphere and was first generated by cyanobacteria (also known as blue-green algae) in the oceans. That is, some scientists believe oxygen did not exist in the atmosphere until early life already existed — via the bacteria consuming carbon dioxide and sunlight and generating oxygen while using the ocean's water to shield it from UV radiation. There is a paradox with this theory, as cyanobacteria would need to be near the surface of the water to receive much sunlight but would need to be several feet or more below the surface to get significant UV protection. E.g., swimmers in the ocean (not wearing sunscreen) are NOT protected from UV radiation from sunlight.

As a geological topic of interest regarding whether early Earth's atmosphere was oxidizing or reducing, extremely durable and stable mineral crystals from a Zirconium ore known as Zircon are known as being almost as old as the earth itself and which are unaffected by environmental conditions that can sometimes affect rock composition. Zircon has been found in what were originally volcanic magnums (known as lavas) which would have contained concentrations of volcanic gases that helped form the early earth atmosphere. The magnums then cooled into rocks that exist today.

Embedded in some Zircon crystals (that were formed as the magnums cooled) is a rare earth metal called cerium, which can be found in two main oxidation states, Ce^{4+} and Ce^{3+}. Ce^{4+} is the more oxidized of the two states. Once the extremely stable Zircon crystals were formed, any

Ce present would <u>not</u> have changed its ratio of Ce^{4+}/Ce^{3+} over time. That is, the Ce^{4+}/Ce^{3+} ratio in lava would have reflected the oxidation state of the atmospheric environment while the magma was hot semi-fluid flowing material in contact with the atmosphere, and would have become permanently fixed once the lava cooled and crystals were formed.

In published research studies, samples of magma rocks known to be at least 4.3 billion years old (which is at least 500 million years longer than the time life is estimated to have existed on earth), many of which were from Australia, were collected with the Ce^{4+}/Ce^{3+} ratio measured. It was determined from this research that the atmospheric composition of early earth (including before life existed) was NOT reducing in nature.[26]

As noted previously, the amino acid building blocks for proteins could only have formed in a reducing atmosphere. Natural phenomena advocates have not been able to produce any geological evidence so far supporting their hypothesis of a reducing atmosphere. They rationalize that the atmosphere was of a reducing nature because that is the only way known in which amino acids could have formed via natural phenomena. So, creationists can point to some geological data that argues against the natural phenomena theory for the origin of life.

Researchers, including Miller, who have repeated the Miller-Urey experiment under the new science based atmospheric assumptions theorizing a more neutral or oxidizing atmosphere, have shown that the new gas mixtures used (or any mixture containing oxygen) do not produce amino acids. It has also been noted by scientists that the concentrated external energy such as lightning and UV radiation that may have helped form "building block" molecules such as amino acids, would also have broken down larger macromolecules essential for life, such as long chain polymers DNA, RNA if they had existed. Also, such non-reducing atmosphere experiments have not produced any nucleotides (the building blocks of RNA and DNA) which are just as critical as amino acids to the process of creating and sustaining life. More on DNA later.

Critics have also pointed out several other aspects of the original Miller-Urey experiment that significantly skewed the results, at least quantitatively. They include:

- Use of a "cold trap" in the experimental apparatus that condensed amino acids as soon as they were formed- rather than leaving them in the gaseous environment (representing the earth's early atmosphere) where they would have been destroyed by the same kind of sparks that helped create them.

- Use of glassware (probably assumed by Miller-Urey to be inert) which catalyzed some of the chemical reactions that took place. Subsequent experiments using the same gas mixture but using Teflon instead of glass for the experimental equipment resulted in a reduced number and concentration of amino acids synthesized.

Several similar experiments by Miller-Urey and others have utilized filtered light (i.e., only long wavelength UV radiation) as the energy source for their experiments. Such filtered light is not representative of sunlight as the short UV wavelength components of sunlight are actually destructive to the chemicals of life. In fact, short wavelength UV is commonly used in commercial sterilization processes.

So, welcome to the world of experimental design. In fairness to scientists, it is challenging to design an experiment and interpret the experimental results that cannot be criticized.

So, the Miller/Urey experiment did show that some living cell building block molecules, such as certain amino acids, can sometimes be created from inorganic chemicals, using gases representing a reducing atmosphere (e.g., no oxygen present). But a reducing atmosphere is no longer thought (by many scientists) to be representative of earth's early atmosphere. Such experiments also produced multiple different structures of individual amino acids when only one structure (known as left handed isomers — described later) are used in living organisms. Such experiments also included other conditions that limit what life forming conclusions could be drawn.[27]

Regarding similar experiments performed in recent years, researchers have tried using various mixes of reducing gases thought to have possibly been present when life began and have succeeded in producing mixtures containing most, although not all, of the amino acids necessary for life. Some of these experiments have apparently also produced some nucleotides (building blocks for RNA and DNA). However, all the gas mixtures used are "reducing" in nature (implying, in part, that no oxygen was present). Many researchers have questioned the assumption of no oxygen in the primordial atmosphere, noting theories (and data from certain space missions) supporting the existence of some oxygen. And, as noted earlier, cerium concentrations in lava rock (originally magnums) suggest the atmosphere was not reducing in nature when life first appeared on earth.

Also, some of the water vapor in the early atmosphere would certainly have reacted with UV radiation from the sun, producing hydrogen and oxygen: i.e., $2 H_2O + \text{ultraviolet light energy} \rightarrow 2 H_2 + O_2$. Ultraviolet light would have been present to enable this reaction as the ozone layer in the upper atmosphere would not have formed yet to block such radiation from reaching the earth's surface.[28]

With this reaction, H_2, a reducing agent, would have quickly diffused into outer space (due to its light weight), with oxygen, an oxidizing agent, remaining in the atmosphere. Gases spewing from volcanic eruptions have also been suggested as a way oxygen might have been present in the atmosphere.

Experiments have verified that when a gaseous mixture is oxidizing in nature (e.g., contains oxygen), organic syntheses effectively turn off. That is, no amino acids or other organic compounds are produced. And, even if they were produced, they would decompose in an oxidizing environment. So, if the early Earth had oxidizing conditions with molecular oxygen present, then spontaneous chemical evolution (leading to life) would have been impossible.[29]

It is of interest that the two Viking missions to Mars in the late 1970s and early 1980s which landed spacecraft on the surface found no evidence of life (i.e., no organic compounds were detected). This was attributed to the presence of UV light that saturates the surface, the dry soil, and the oxidizing nature of the soil chemistry. Also of interest is that the Martian atmosphere contains about 0.2% oxygen and no hydrogen, so is slightly "oxidizing" and not "reducing." Recall that scientist's experiments using an "oxidizing" mix of gases have not produced biomolecules.

Subsequent experiments, such as using the spectrographs on the Hubble telescope, have shown that the atmospheres of two moons (Euopa and Ganymede) of Jupiter contain some oxygen. The Hubble telescope has also apparently detected the presence of oxygen in the atmosphere of the hot exoplanet, Osiris (a planet in the constellation Pegasus). The point is that there are other ways that oxygen can appear in an atmosphere than the usual cause familiar to most people- that being photosynthesis in living plants. So, the presence of some oxygen in an atmosphere does not require that life be present.

Further, oxygen must have been present in the early atmosphere to enable the creation of the ozone layer — with the ozone layer required to block UV radiation so that life could exist. That is, Ozone = O_3.

Ozone is created when molecular oxygen (O_2) is exposed to an electrical field or ultraviolet (UV) light (such as exists with sunlight). This causes some of the O_2 molecules to split into individual oxygen atoms (O). The free oxygen atoms combine with O_2 molecules to form ozone, i.e., $O + O_2 \rightarrow O_3$. So, oxygen must have been present in the atmosphere in order for the protective ozone layer to form, thus suggesting an oxidizing environment.

Other experiments in aqueous solutions have shown that while high energy sources are needed to produce most of the amino acids and certain other biochemicals needed for life (in reducing environments), that such energy sources do not necessarily have to be lightning, UV radiation, or heat but, rather, can be high energy chemicals (e.g., formaldehyde, cyanates) that might be present in aqueous environments, (e.g, ponds, oceans).[30]

Such experiments have encouraged some scientists to continue the quest to show life could have started via chemical evolution. However, there are many barriers with their mission, one being that high energy sources are often more destructive than constructive, especially for larger molecules, sort of like a "bull in a china shop." For example, UV light is lethal to living organisms but is still considered by many chemical evolution scientists to be essential (as an energy source) to the origin of life.

Summarizing, it appears that life originating from natural phenomena was highly unlikely regardless of the various proposed theories regarding the composition of early earth's atmosphere. E.g., regarding oxygen:

If oxygen did not exist in the early atmosphere:

- no protective ozone layer develops

- UV radiation from the sun destroys any large bio-molecules that may have formed such as proteins and/or nucleotides needed for RNA, DNA.
- result: no life (via natural phenomena)

If oxygen existed in the early atmosphere:
- no chemical reactions occur creating amino acids (the building blocks of proteins)
- result: no life (via natural phenomena)

While experiments such as those conducted by Miller-Urey are of some interest, the gap between a soup containing a few of life's building block molecules, such as amino acids, and a living cell is huge, in fact, astronomical. So, it is puzzling that a well-meaning but misguided experiment (that produced less than 1/100 of 1% of the different kinds of components needed for a living cell), while using unrealistic experimental conditions and raw material composition, has received so much publicity. Regardless, as aerospace engineer Dr. Wernher Von Braun stated: "One good test is worth a thousand expert opinions."

So, for scientists advocating the natural phenomena theory for the creation of life, they have an astronomically long way to go to prove their hypothesis —so they need to keep those experiments going.

The book, *The Mystery of Life's Origin, the Continuing Controversy*, in its conclusion on this topic and in acknowledging past interest in the Miller-Urey experiment and its inclusion in popular biology textbooks, states: "For good scientific reasons, the Miller-Urey experiment is now widely thought to be irrelevant to the Origin of Life on Earth discussion."

More later on other topics that contribute to the astronomical gap between primordial soups and actual life.

CHAPTER 4

Biogenesis and Abiogenesis

Objective: Describe Biogenesis and Abiogenesis— which are conflicting theories, both developed by scientists, regarding the origin of life.

relevant perspective on the "origin of life" comes from the field known as Biogenesis, the "father" of whom is Louis Pasteur.

Note: Pasteur was a French chemist and microbiologist renowned for his discoveries regarding vaccination and pasteurization. He is regarded as one of the founders of modern bacteriology and as the father of microbiology. Biogenesis argues against natural phenomena as the origin of life. In 1851, in one of Pasteur's experiments (diagrammed below), he boiled (i.e., sterilized) nutrient rich beef broth, initially sealing flask exit tubes and later, opening up some of the exit tubes. Sterilizing, of course, would have killed any and all microorganisms in the broth and exit tubes.

Subsequent to sterilization, with the experiment lasting about a year, no microbes (i.e., living organisms) appeared in the sealed containers (despite the rich mix of available nutrients in the beef broth), but they did appear in the unsealed containers (in which the tube contents were exposed to the surrounding atmosphere –which contained microbes).

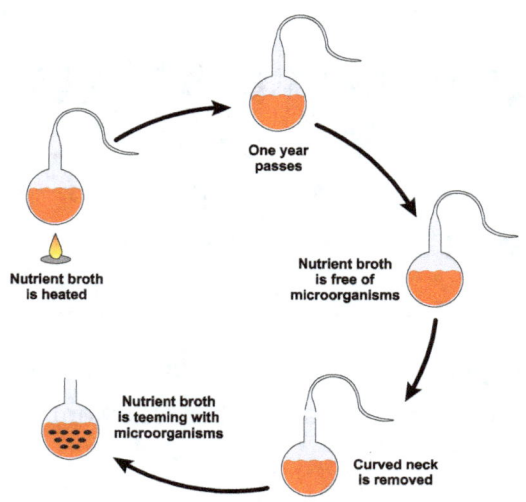

One year
passes

Nutrient broth
is heated

Nutrient broth
is free of
microorganisms

Nutrient broth
is teeming with
microorganisms

Curved neck
is removed

FIGURE 10: PASTEUR'S FAMOUS BIOGENESIS EXPERIMENT

The conclusion of Biogenesis, in general, and Pasteur's experiment, in particular, is that "life does not spontaneously arise from non-living material." In other words, if "all life is from life," then all new cells are from existing cells.

> **Note:** This is a key experiment arguing against some scientist's claims that life arose from non-biological sources.

The biogenesis conclusion that "all life is from life" is one of the central themes of "cell theory." So, where did the first cell come from? Cell theory doesn't say.

A popular theory subscribed to by naturalists is Abiogenesis. Abiogenesis is the study of the creation of organic molecules by forces other than living organisms and includes the theory that life arose on Earth via

spontaneous natural means from non-living matter. The original experiment by Miller and Urey, described in Chapter 3 of this book, helped
spawn the field of Abiogenesis.

So, it seems that the life science disciplines of Biogenesis (i.e., "all life
is from life") and Abiogenesis (i.e., "life arising from non-living matter"),
each of which was developed by scientists, are in direct conflict.

CHAPTER 5

Other Concerns with the Natural Phenomena Theory that Pertain to the Origin of Life

Objective: Summarize other troubling aspects of the natural phenomena theory regarding the origin of life. Examine different kinds of locations where some scientists believe life might have begun.

There are additional perspectives that some authors have offered that challenge natural phenomena as being responsible for the origin of life. For example, water is the liquid of life —whereas carbon provides the framework for all biomolecules. Many of the key chemical steps in assembling carbon-based biomolecules don't naturally work well in water unless catalyst assistance is available. That is, most organic chemical reactions normally work best in solvents other than water (and biomolecules are organic compounds).

Another issue regarding water is that it causes "hydrolysis" which is the breakdown of certain substances by water. So, even if large polypeptides (i.e., proteins) and other biopolymers had somehow formed in the prebiotic "water based" soup, they would have then been subject to hydrolysis. Some amino acids might also have been affected.

The literature is unclear as to how important hydrolysis is regarding biomolecules, with some authors indicating it to be a factor and others noting the reaction is so slow (under normal cell pH and other conditions) as to be negligible. Also, if water itself was not of enough concern, polypeptides and polynucleotides (e.g., DNA, RNA), once formed, would have been vulnerable to degradation by chemical interaction with a variety of other substances in the ocean/lake/pond waters such as amines, aldehydes, ketones, reducing sugars, and/or carboxylic acids.[31]

Given the apparent ability for biomolecules to form and exist in living cells (in a water environment) and remain stable for significant periods of time (months or even years), it would appear that some of the above listed biomolecule reaction concerns, while possible, are negligible.

> **Note:** Many, if not most, biochemical reaction rates (whether associated with synthesis or degradation) are determined by the availability of specific catalysts.

Anyway, to many scientists, the great mystery of life's origins lies in the gap between simple organic molecules and primitive cells.[32] So, much has yet to be explained and proven (if natural phenomena is indeed the answer).

Dr. Michael Denton, Ph.D., Senior Fellow at the Discovery Institute's Center for Science and Culture, has stated, "Considering the way the "prebiotic soup" is referred to in so many discussions of the origin of life as an already established reality, it comes as something of a shock to realize that there is absolutely no positive evidence for its existence."[33]

In addition to the challenge of explaining "How" life began, the scientific community is also split as to "Where" life began. Theories range from life starting 1) in a pond (i.e., in a primordial aqueous soup containing organic molecules) with energy supplied by lightning and/or UV rays from sunlight, to 2) in the ocean bottom (by hydrothermal vents) to avoid destructive UV rays and lightning — with energy provided by sulfur compounds coming from the vents, to 3) on land based rock, clay, or mineral surfaces, to 4) from introduction to earth (at least for many of life's chemical building blocks) via meteorites and/or comets. As with the "how," only unproven theories exist as to the "where."

FIGURE 11: POSSIBLE VIEW OF EARTH'S SURFACE BEFORE LIFE BEGAN

FIGURE 12: A WATERY AREA PERHAPS CONTAINING A PRIMORDIAL SOUP OF CHEMICAL COMPOUNDS

FIGURE 13: HYDROTHERMAL VENTS NEAR AN OCEAN BOTTOM

FIGURE 14: LIFE BEGINNING FROM DEPOSITS ON ROCKS, CLAY, AND/OR MINERALS

FIGURE 15: DID METEORITES BRING LIFE'S BUILDING BLOCKS TO EARTH?

Emergence

A few scientists interested in the "origin of life" have coined a new phrase, "emergence", to characterize what some of them are working on. Emergence is defined as the process by which simple systems of many interacting particles spontaneously become more complex. The gist of "emergence" seems to be that energy levels within a certain range (not too low and not too high) can sometimes create patterns in mass collections of interacting particles (or objects) when subjected to an energy flow, thereby creating something more complex than exists with individual particles (or objects).

Examples cited are the "arms" of spiral galaxies in the universe, collections of sand particles (with energy from winds) forming sand dunes, collections of neurons in the brain causing a conscious state, and particles in liquids forming highly structured lattices as they cool and crystallize. One can question if these are relevant analogies that at least partly explain how chemical components could have come together to create life or, e.g., explain man's consciousness.

Scientists working on "emergence" theories hope that a new quantitative law of physics will eventually be discovered to explain such phenomena, but no such law exists yet, or is even close. Regardless, the gap between understanding / quantifying "emergence" and its possible role in the "origin of life" is enormous.

The Bible

The literature dealing with the "origin of life" often cites Bible verses to support the "creation" theory. The Bible, of course, is the most published and distributed book in the history of the world and is widely accepted as factual by hundreds of millions of people. However, the

Bible's coverage of creation and the beginning of mankind is limited, utilizing only a few verses of text and containing little detail — therefore opening the door for various interpretations and for other theories to be hypothesized. The Bible's Genesis even has some hard to explain verses, such as ones indicating light (including night vs. day) existing (creation day 1) before the stars (the source of most light) were formed (creation day 4).

> **Note:** While the bible does not provide much detail on the origin of life, it does repeatedly note (by different authors) that God created heaven and the earth. As noted earlier, Genesis 1 of the Old Testament, presumably written by Moses, states: "In the beginning God created the heaven and the earth." In Acts 17, in the New Testament, the apostle St. Paul writes: "…God, who made the world and everything in it, …"

The Bible teaches that "God made man in His image". It is known that human's creative powers have been needed and used from the earliest times of human existence (e.g., to develop hunting and building tools) to current times as humans have developed and constructed the buildings, machines and complex physical systems that are used in people's daily lives. If humans are created in God's image, isn't it reasonable to believe that a God exists, is smarter than humans, and has been involved in creating the universe and those things in it that are beyond what humans have been capable of creating, including life itself?

CHAPTER 6

The Structure and Function of Living Cells: Overview

Objective: Examine the general biological and chemical structure and function of living cells so that readers might consider if such structure and function could have occurred via natural phenomena. Consider the complexity of cells from multiple perspectives, including the construction of proteins from amino acids, the construction and function of DNA, the complex metabolism of cells, the need for the entirety of cellular machinery to be contained in a semi-permeable membrane, and the requirement for most cells to create copies of themselves.

This chapter is intended to give the reader an appreciation of the complexity of living cells and not necessarily to have readers understand all the details.

The simplest self-contained form of life is the living cell. Some readers may think a living cell is a relatively simple thing — especially if they have never completed a biology course in school. And, as noted earlier, scientists thought so when Darwin came up with his theory of evolution in the mid 1800s.

However, each living cell is actually very complex. Further, as complex as a single cell is, even far greater complexity exists when considering an entire organism such as the human body. The human body contains tens of trillions of cells contributing to about 200 different types of tissues (nerve, muscle, etc.), all integrated and working together for the growth, maintenance, and reproductive needs of the organism—as well as, at least in humans, somehow supporting emotions, consciousness, morality, and spirituality. Regardless, taking first things first, the structure of and operations within a single cell is the focus of the next part of this book. After all, a necessary condition for an organism to be viable is for the building block cells of the organism to be viable.

A simplified view of the structure of a simple cell (such as bacteria) and that of a more advanced cell (representing cells in plants and animals) are shown below:

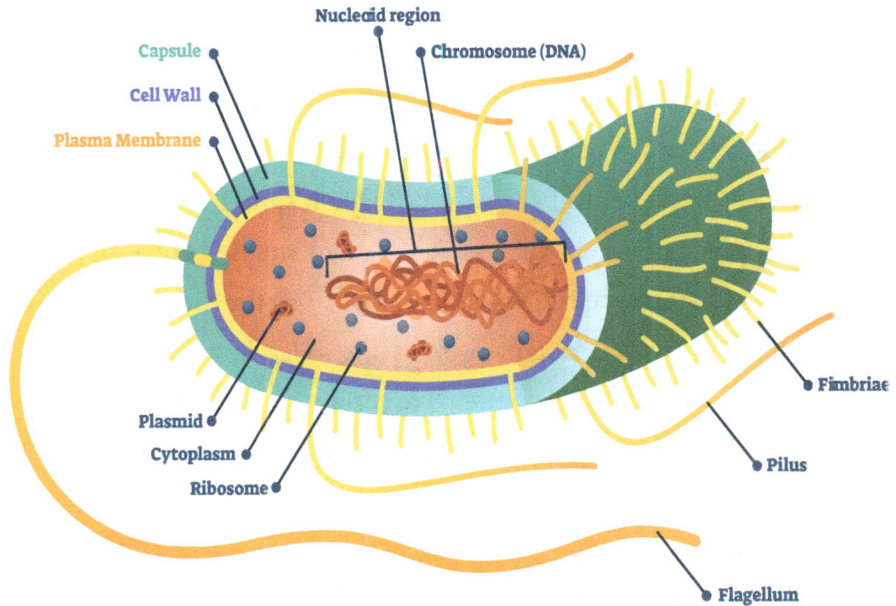

FIGURE 16: BASIC STRUCTURE OF (PROKARYOTIC) SINGLE CELL ORGANISMS (TYPICAL OF BACTERIA)

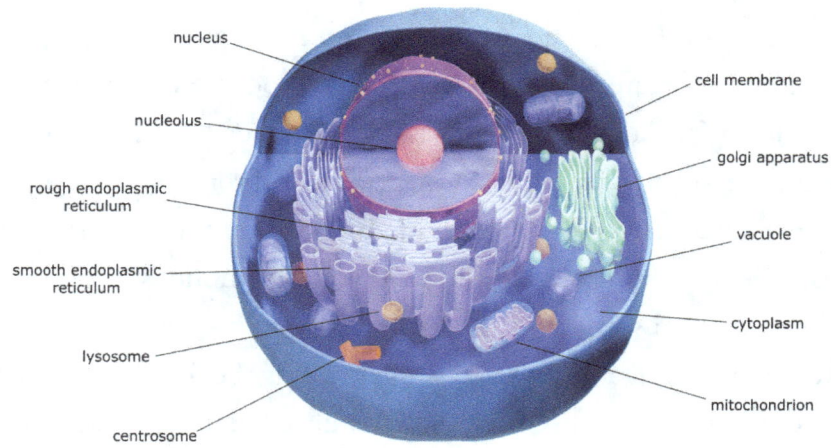

nucleus

cell membrane

nucleolus

golgi apparatus

rough endoplasmic
reticulum

vacuole

smooth endoplasmic
reticulum

cytoplasm

lysosome

mitochondrion

centrosome

FIGURE 17: BASIC STRUCTURE OF MORE ADVANCED , EUKARYOTIC, TYPES OF CELLS
(TYPICAL OF CELLS IN PLANTS AND ANIMALS)

Note: Much of the discussion in the rest of this chapter applies to both kinds of cells —as both kinds of cells have the same requirements for life (i.e., growth, maintenance, reproduction). Some chapter content applies only to the more advanced eukaryotic types of cells which are typical of cells in plants and animals. It is worth summarizing the difference between eukaryotic cells and simpler unicellular prokaryotic cells, such as bacteria, since the first examples of life were probably prokaryotic cells.

Prokaryotic cells do not have certain structures within the cell such as Golgi apparatus, lysosomes, or mitochondria that exist in eukaryotic cells and they do not have a well-defined nucleus containing linear DNA. Rather they have a less well-defined nucleus (called a nucleoid), with DNA consisting of a single chromosome circular molecule (meaning it does not have a beginning and end but is continuous) in the cell's nucleoid region plus smaller circular DNA ringlets dispersed in the cytoplasm known as plasmids.

All cells have a highly organized and complex structure. For plant and animal cells, the living cell has aptly been likened to a "human designed and built" manufacturing factory, complete with a boundary fence (the cell wall), gates, docking bays and security systems, entry facilities for raw materials, shipping facilities for finished products, internal transport systems, power plants (mitochondria), waste disposal plants (proteasomes), machines for manufacturing proteins (ribosomes), an army of workers with many different skills (enzymes), messengers (mRNA), stock-pickers (tRNA), and blueprints (DNA).[34]

There are lots of molecules within a cell, such as the millions of components (nucleotides) that make up DNA and RNA, the thousands of different proteins, and the thousands of different carbohydrate molecules (e.g., cellulose for plants), as well as lipids and other molecules that make up portions of cell membranes/walls.

A few (not all) of the cellular components shown in the above figures are briefly described further in this book, primarily to "paint a picture" of the high level of complexity of cells. It is to the credit of scientists that so much has been learned about the structure and function of cells in recent decades. Regardless, new discoveries about cells serve, in part, to reveal an even greater complexity of cell structure than what was previously known.

Such discoveries are analogous to those occurring in other science disciplines (e.g., atomic physics). For example, for a long time, it was thought that the smallest particles in nature were the electrons, neutrons, and protons that make up atoms. It has since been learned that atoms are more complex than this, with their structure including quarks and other subatomic particles.

So it is with cells. Cells are NOT just a collection of various compounds in aqueous solution surrounded by a membrane. Like an automobile's drive train (although far more complex), cells are made up of various interacting components and substructures, all of which must be working for the cell to survive. Each major component of cells (e.g., nucleus, ribosome, cell membrane) is, itself, a complicated structure. The more complexity that scientists reveal about cells, the more one wonders how such complexity could have occurred by natural events, even if over a long period of time.

If some of the labels of cellular components in a living cell (as shown in Fig. 17) are not immediately recognized by readers, it may help to know that most components are roughly analogous to structures in the human body. For example, the cell has a control and information center (analogous to the brain) which is the DNA in a cell's nucleus. The cell has a cell membrane perimeter analogous to the body's skin. Many of the cell's other components (sometimes called organelles) are analogous to the body's organs, as each provides unique specialized functionality. Of course, while anatomical details differ, the function of cells and complex organisms have much in common; they each must take in nutrients, convert them to useful energy, eliminate waste products, grow, maintain and control operations, and reproduce.

The first forms of life, undoubtedly, were single celled microorganisms. Such cells were similar to, and might have included, *E. coli* bacteria which many readers will recognize as one of the microorganisms existing in a mammal's digestive track intestines (to help with digestion). *E. coli* bacteria are noted here as they have been studied extensively by scientists. They are used in the commercial production of several prod-

ucts, such as insulin, with much known about their anatomy (i.e., structure) and physiology.

Every living cell, including the simplest single-cell organisms, is
complex and complete. No cells have been discovered in fossils in some
stage of partial development from non-living matter. So, from this perspective, there is no physical evidence of the evolution of non-living
matter into living cells.

Most cells have certain structures and components in common. That
is, most cells contain DNA, RNA, ribosomes (where proteins are made),
a semi-permeable cell membrane/wall, an internal mixture of many proteins and other molecules, a mechanism to acquire and store chemical
energy, a mechanism to maintain metabolism, and a mechanism for
growth and reproduction.

Regarding cells that make up plants and animals, most cells contain mitochondria which are called the "powerhouse" of the cell since
they generate most of the cell's supply of ATP (adenosine triphosphate)
which is the source of chemical energy needed by cells. Most such cells
also contain other structures (e.g., Golgi apparatus, endoplasmic reticulum) that are beyond the scope of this book to describe. Readers should
just be aware that the few topics that are summarized in this book are
just the tip of the iceberg in terms of the known complexity of cells.

Each living cell is a tiny container of thousands of different chemicals
and water wrapped in a membrane. At least 11 elements are contained in
every cell, all of which are necessary for life (carbon, oxygen, hydrogen,
nitrogen, calcium phosphorus, potassium, sodium, sulfur, chlorine, and
magnesium). Additional elements exist in certain kinds of cells, such as
iron needed for hemoglobin in red blood cells.

Each cell in an organism (with a few exceptions) contains all the
genetic information necessary to manufacture whatever organism (e g.,

human) that it is part of. Pause for a moment and ponder this statement. That means that the DNA in a muscle cell also contains the blueprint for making nerve, cartilage, and other kinds of cells. What would cause the DNA in an individual cell to have evolved to where it not only enables replication of similar cells, but, even more amazingly, contains the necessary information for the manufacture of cells unlike itself?

> **Note:** Human stem cells not only contain the code for making almost any kind of cell in the body, but are capable themselves of reproducing to become almost any of the different kinds of cells in the body.

At any given time, each cell is doing thousands of routine jobs, such as capturing and using energy, manufacturing proteins (and other chemical compounds) and responding to environmental cues. For multicellular organisms, different cell types also have special duties, such as building skin or bone, generating hormones or making antibodies.

The complexity of individual cells is briefly discussed in the following several pages, summarizing a few of the many key functions of a cell. It begins with a discussion of the structure of proteins and then summarizes 3 key requirements for cell life, all of which require proteins: 1) protein manufacture and cell reproduction, 2) cell metabolism, and 3) cell membranes. The content should be helpful to readers in deciding for themselves how likely such complexity might have developed due to processes such as random environmental events, mutations, and natural selection.

Proteins

General Characteristics. Each human cell contains many different kinds of molecules, such as carbohydrates, lipids, and proteins. In considering proteins, up to 20,000 different kinds of proteins exist in a cell with the total number of protein molecules in a cell typically approaching 2 million or more.

Proteins are molecules that perform a host of essential tasks. Without proteins, cellular metabolism (i.e., maintenance) cannot exist, and cells cannot grow and multiply. For animals, hearts cannot pump, lungs cannot breathe, food cannot be digested, and blood in open wounds cannot clot.

Proteins provide cellular structural material, control cell growth and metabolism, and include hundreds of enzymes that help carry out activities within the cell by catalyzing chemical reactions that would otherwise occur only slowly or not at all.[35]

Each protein is essentially a long chain (or multiple linked chains) of amino acids, strung together in a specific sequence. There are 20 different kinds of amino acids specified by the standard genetic code as protein building blocks (with an average molecular weight of 110 each).

Note: There are a couple of additional amino acids that can be used for special applications).

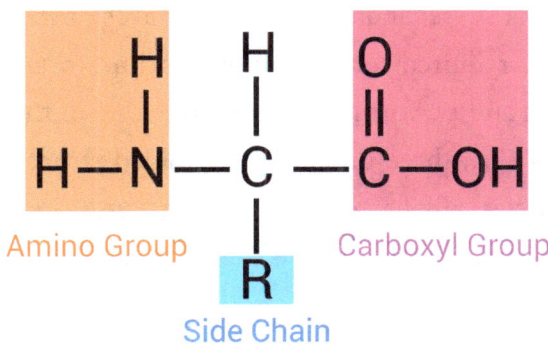

FIGURE 18: THE CHEMICAL MAKEUP OF A TYPICAL AMINO ACID

> **Note:** Amino acids differ from one another primarily in having a different side chain "R"; e.g., for the amino acid, alanine, the side chain (R) is the methyl group, CH_3.

The number of amino acid "building block" molecules in each protein molecule ranges from 50-2000. However, the number of amino acid molecules in most proteins is in a narrower range of between 300 and 400. So, the molecular weight of an average protein molecule is about 38,000 (compared to the molecular weight of a hydrogen atom = 1, or a carbon atom = 12).

The primary structure of protein chains (also known as peptide chains) is created by connecting amino acid molecules together with each connection (i.e., bond) consisting of an amino (NH_2) group of one amino acid linked to the carboxyl (COOH) group of the next amino acid. As each bond is formed, a molecule of water is formed and released.

There are many aspects of proteins that contribute to their complexity and summarizing all of them is beyond the scope of this book. So, only two aspects will be mentioned as examples: 1) the tertiary (i.e., folded) structure of proteins and 2) their isomer specificity.

Protein Tertiary Structure. Each protein molecule, by itself, has an important primary, secondary, tertiary, and sometimes quaternary structure that is required for the protein to link to and/or react with other molecules and perform their function. One of these is the highly specific tertiary (i.e., three dimensional) folded structure required for the protein to physically fit into biological receptor sites (like a ball and socket union — or a uniquely shaped key needed to operate a particular lock) and do its job.

For example, there are receptor sites on the surface of most human cells for which only the unique 3 dimensional structure of insulin protein molecules can fit onto — thus allowing insulin to do its job in enabling glucose to enter the cells.

Mechanism of action of the enzyme. Key and lock hypothesis

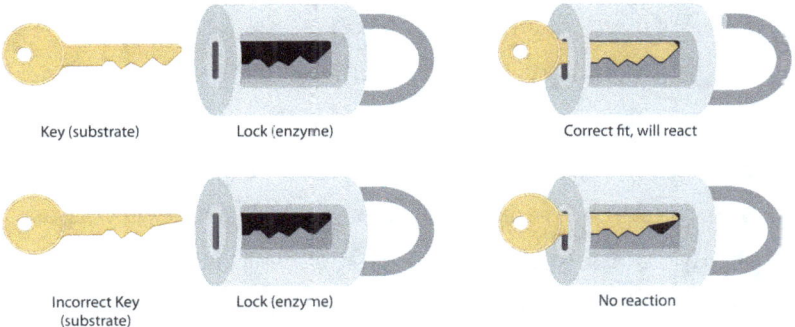

Key (substrate) Lock (enzyme) Correct fit, will react

Incorrect Key (substrate) Lock (enzyme) No reaction

**FIGURE 19: PROTEIN LOCK AND KEY ANALOGY
A REQUIREMENT TO ALLOW FOR PROTEIN REACTION AND FUNCTION.**

The three-dimensional folded protein molecule structure is determined largely by 1) the sequence of the amino acids used as building blocks in constructing the protein and 2) the internal linkages (i.e., bonds) connecting certain amino acids within the protein molecule. That is, a protein is thought to be generated such that it initially looks like a string (or multiple strings) of different sized and shaped beads — with each bead being an amino acid.

These strings then fold (i.e. compress) upon one another into a specific compact three-dimensional structure, held together by internal bonding between some (or all) of the beads (i.e., amino acids). The strongest of these links, for proteins containing the amino acid cysteine, are covalent disulfide bonds attracting cysteine molecules to one another. The remaining links that help provide three-dimensional structure for proteins are mostly weaker hydrogen bonds in which most or all amino acids participate.

Note: While details are beyond the scope of this book, it is difficult to imagine how a newly formed linear strand of 350 amino acids (which is an average sized protein) inside a cell will almost always fold into the correct specific 3-dimensional configuration.

As an analogy, imagine taking a string of beads several feet long and wadding it up into a small volume. Its 3-dimensional structure would probably be a tangled mess and be a visually different tangled mess every time the operation was repeated. That doesn't happen when a linear protein molecule folds inside a living cell.

Most proteins cannot spontaneously fold themselves and, even if they could, the large number of attractive and repulsive forces in play between the many atoms involved would suggest multiple folded structures would be possible. Yet that is not what happens inside cells. It turns out that protein folding is not a random or spontaneous process but a precisely orchestrated event.

Proteins undergo a series of modifications and encounters with hundreds of molecular chaperones and folding enzymes that together assist in proper folding of the peptide, normally accomplished in ribosomes in the endoplasmic reticulum portions of cells. Chaperones and enzymes are themselves properly folded proteins.

So, how would such a complex protein folding control mechanism have come to be (by natural phenomena), for each of the thousands of different proteins in a cell? That is, to properly fold a protein requires the existence of other proteins. So how did the first proteins come to be?

Protein quaternary structure regards those protein molecules that consist of multiple amino acid chains, bonded together. For example, as will be shown in the next section, an insulin molecule consists of two differently constituted amino acid chains, bonded together. A hemoglobin molecule consists of four different amino acid chains bonded together.

Example Protein: Insulin. An example of a representative protein molecule, albeit a very small one, is the well-known protein, insulin.

> **Note:** While used as an example of a protein, insulin is not made by most human cells. It is made by the beta cells in the pancreas gland and used in conjunction with liver and certain other cells for the purpose of controlling blood sugar level. However, insulin is a typical protein in many respects and is one of the best known and widely studied proteins.

A molecule of insulin consists of a sequence of 51 amino acids (split into 2 interconnected peptide chains), and has a molecular weight of 5,808. Its chemical formula is $C_{257}H_{383}N_{65}O_{77}S_6$. So, each molecule of insulin contains a highly specific 3-dimensional configuration/structure of 257 carbon atoms, 383 hydrogen atoms, 65 nitrogen atoms, 77 oxygen atoms, and 6 sulfur atoms. The sulfur atoms (from the amino acid cysteine) are, in part, involved in the internal linking (i.e., disulfide bonds) of the 2 peptide chains which is important in maintaining the required 3-dimensional shape of the insulin molecule.

There are many insulin molecules within a pancreas beta cell — with the insulin molecules only one of thousands of different kinds of proteins that exist in beta cells. Insulin molecules are excreted by the beta cells in the pancreas with most molecules then existing in the blood stream (or attaching to receptors of other cells) where they help control blood glucose concentration.

A diagram of an insulin molecule is shown below. It shows the two different strands of amino acid sequences, with the two strands bonded together. The diagram is shown in 2 dimensions, but the actual molecule is three dimensional.

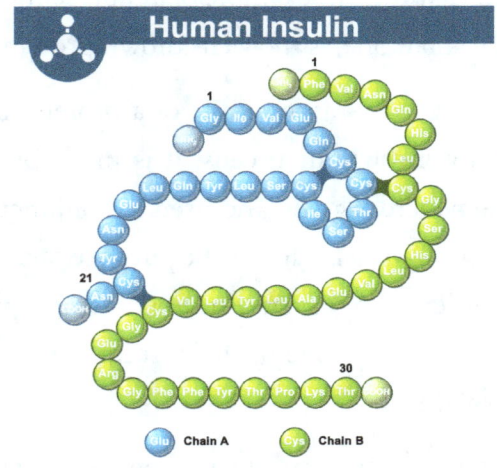

FIGURE 20: A HUMAN INSULIN MOLECULE (IN TWO DIMENSIONS)

The individual circles in the above diagram represent different amino acids and so represent different protein building block molecules of different physical sizes (even though, for diagram simplicity, the beads in the above diagram are all the same size). Note the three places in the image where linkages between amino acids are shown. These are strong disulfide bonds that link cysteine amino acid molecules together. As with most proteins, there are many other linkages (mostly weaker hydrogen bonds) in an insulin molecule helping to hold its 3-dimensional structure together.

So, if any of the six cysteine amino acid molecules in the insulin molecule were to be replaced with any of the other 19 types of amino acids used by organisms, at least one of the disulfide bonds would disappear which would significantly alter the three-dimensional structure of the molecule, undoubtedly resulting in the molecule's loss of functionality.

> **Note:** Scientists have been able to make an amino acid swap with one of the non-cysteine amino acids in the insulin molecule which has allowed the molecule to retain its approximate basic three dimensional folded structure while resulting in a modification in functionality (e.g., creating a beneficial "longer lasting" form of insulin).

The actual 3-dimensional insulin molecule looks approximately as follows:

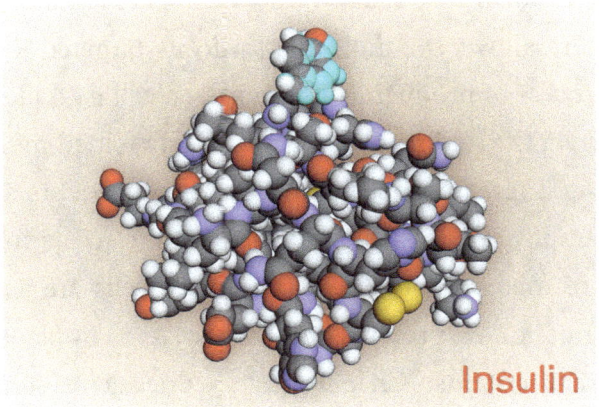

Insulin

FIGURE 21: 3-DIMENSIONAL VIEW OF A HUMAN INSULIN MOLECULE

To add perspective regarding protein structure from the 2020-22 corona virus pandemic, the following figure illustrates how antibodies (which are proteins) attach to a virus (e.g., the Covid-19 virus). The sites of attachment conform to the Lock and Key principle discussed previously and shown in Figure 19.

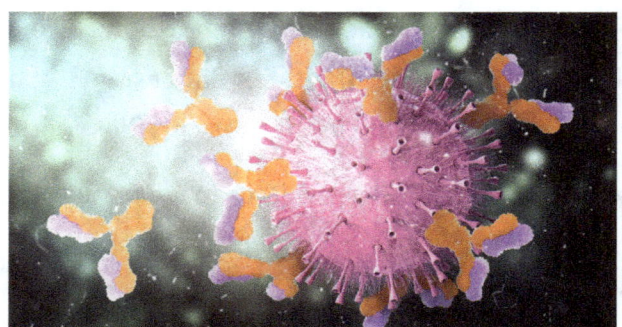

FIGURE 22: ANTIBODIES (WHICH ARE PROTEINS) ATTACHING TO VIRUS RECEPTOR SITES
NOTE: RECEPTORS ARE ALSO PROTEINS, SHOWN AS TUBULES ON THE VIRUS SURFACE.

Improper Protein Folding. When microorganisms manufacture a protein (or its precursor), most of the molecules that are produced fold into their correct three-dimensional structure, but some molecules do not. Often, different three-dimensional structures of a protein are theoretically possible to form but, usually, only one specific three-dimensional folded structure allows the molecule to do its targeted job. Therefore, part of the process of producing a protein in a cell is to identify and get rid of (or correct the folding of) any improperly folded molecules.

Generating a few improperly folded proteins occurs sometimes in cells. For example, one mis-folded protein, gone rogue inside the cell, is the cause of mad cow disease. Fortunately, cells are accustomed to coping with mis-folded proteins and have systems in place to refold or destroy aberrant proteins. "Chaperones" are one such system. Appropriately named, chaperones accompany proteins through the folding process, improving a protein's chances of folding properly and even allowing some mis-folded proteins the opportunity to refold. Interestingly, chaperones are proteins themselves.

Another line of cell defense against mis-folded proteins involves a "proteasome." If mis-folded proteins linger in the cell, they will be targeted for destruction by this molecular machine, which chews up proteins and spits them out as small fragments of amino acids. The proteasome is like a recycling center, allowing the cell to reuse amino acids to make more proteins. The proteasome itself is not just one protein but many proteins acting together.

> **Note:** Improperly folded proteins can also occur when final folding occurs outside the cell walls from the protein's precursor that is made inside cells, such as occurs in some commercial processes. For example, one commercial process for making insulin uses *E. coli* bacteria cells to make proinsulin (a precursor for insulin). Proinsulin is then recovered from the cells and converted to and folded into the final insulin form during subsequent chemical manufacturing steps occurring in standard chemical reactors (i.e., external to cells). In such cases, the cells are not available to identify and correct improperly folded insulin product molecules, so other man developed methods are needed.

Thus, part of the remarkable complexity of cells is their ability to diagnose many of their own problems, such as improperly folded proteins, and fix them. How did cells ever acquire the ability to do this? In thinking about the most advanced non-biological systems in society today, there are several that can <u>diagnose</u> a few of their own problems — with some diagnoses even triggering an automated action step, e.g., starting up a backup generator after an electric utility power failure. Any such capabilities were developed by the creative intelligence of humans. Regardless, one is hard pressed to name even a single non-biological system that can automatically <u>fix</u> its own problems. Using the above example, turning on a backup generator doesn't fix what originally caused the power failure. More on the body's repair mechanisms later.

Isomers. Another baffling aspect of protein make-up is in their selectivity of the correct "version" (i.e., structure) of each amino acid "building block" utilized. The different versions of a specific amino acid have the same chemical composition, just different structural orientations, known as isomers (or sometimes stereoisomers).

> **Note:** Many sugars and other non-protein molecules can also exist as isomers. For example, the sugar molecule glucose can have 16 different structure orientations.

The topic of isomers is complex and so with respect to this book, only the subset of these isomers that are mirror images of one another will be discussed. The mirror-image isomers are known as being either left handed (L) or right handed (D) versions of the molecule. These isomers get their description as left handed or right handed based on an analogy to gloves, in which the glove for the left hand (L) is the mirror image of the glove for the right hand (D). A similar analogy exists with left vs. right handed golf clubs. Chemical isomers are identical with respect to composition, size, and most other physical properties but cannot be superimposed on one another.

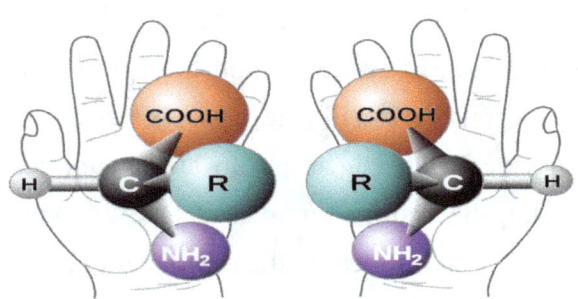

FIGURE 23: VIEW OF LEFT- AND RIGHT-HANDED AMINO ACID ISOMERS

While not an amino acid, it is of interest that the two most common isomers of glucose (a sugar) even taste the same — suggesting that the mirror image difference of the two glucose isomers do not affect the two isomer's ability to link up to receptor sites on the tongue to trigger a "sweetness" taste. Interestingly, the L-glucose isomer (which has zero calories) has been explored as a sugar substitute for the far more common high calorie D-glucose isomer. I.e., the L-glucose isomer provides no nourishment since it cannot react with and be broken down by the body's cellular enzymes. That is, catalysts in the body know the difference between different mirror image isomers of a target molecule, perhaps reacting with one but not the other.

The distinction between isomers is relevant because all known non-life chemical synthesis processes that generate amino acids yield a mix of mirror-image isomers (usually about a 50: 50 mix). However, it turns out that proteins in most cells consist exclusively of left-handed amino acid isomers. How can this be?

Said another way, at the molecular level, life is one-handed, (i.e., homochiral) even though non-life is two-handed and shows no preference between left- and right-handed isomers. Without isomer purity (i.e., use of only left-handed isomeric forms of amino acids), most proteins would not function properly.

All known laboratory experiments involving simulated prebiotic soups produce a random distribution of left and right-handed isomers of molecules and not just one specific isomer (i.e., of molecules where multiple isomers can exist). For example, a common non-biological commercial method of producing amino acids is the Strecker synthesis method — which produces both isomeric forms.

Despite all that chemists know about synthesizing chemicals, they have yet to line up amino acid molecules (contained in multi-isomer mixtures) that have only the correct "left glove handedness" and assembling them together in the correct sequence to make a desired protein. So, somehow, whether generated in the first cells or having evolved over time, living cells know how to selectively use (or reject) specific isomeric forms of amino acids. So, on the ribosomes of cells (used for making proteins), only the L-isomer form of amino acids is selected and used.

Also, while the L form of amino acid isomers may be predominate in today's living cells (and in the diets that provide nutrients to cells), they would not have been predominate in creating life's first living cells, as prebiotic soups would have contained both mirror image isomers, presumably in roughly equal amounts.

So, it is apparently not complex enough that metabolic pathways involve many thousands of different compounds (as shown in the next section). They also require use of specific isomers for the many compounds that have isomers (including most amino acids and also glucose — with glucose existing and used in all cells).

The development of isomer differentiation capability in cells is unknown, but some causes can be eliminated. I.e., isomer differentiation has nothing to do with physical forces on earth (e.g., gravity, electromagnetism) as such physical forces act equally on both L and D (i.e., mirror image) isomers of a molecule. Also, differentiation has nothing to do with bonding between amino acids. That is, amino acid D-L links form just as easily as do D-D or L-L links. Isomer purity is a characteristic of life. The mystery is how this selectivity emerged from the random prebiotic world.[36]

So, the world of proteins inside a cell is clearly very complex.

DNA and RNA
Structure, Making Proteins, and Cell Division

Even more complex than the presence and reactions of proteins inside a cell are the instructions controlling if, when, and how a cell manufactures individual proteins. As an analogy, think about what is involved in manufacturing a car. A car isn't assembled just because all the right parts are nearby. A car's manufacture needs to:

1) be scheduled

2) have the right parts available at the right place at the right time

3) have assembly equipment (including robotics and automated welders) available and in working condition

4) have thousands of approved procedures in place that specify all aspects of manufacturing the car

5) have hundreds of validated computer programs executing

6) have trained assembly workers at their workstations

7) have additional workers present to perform quality control inspections (according to defined instructions), fix problems, and recycle "lemons" for rework.

Many of these activities are coordinated through a company's centralized plant computer system known in the IT (Information Technology) world as a MES (Manufacturing Execution System).

Protein manufacture has its own "MES" system, known as DNA. Instructions in DNA include:

1) the specification of the correct sequence of amino acids for the different kinds of proteins

2) the control of when manufacturing of specific proteins is to start and stop

3) specification and control of catalyst manufacturing to support protein assembly

4) assembly of separate proteins (e.g., chaperones to help identify and recycle improperly folded proteins) to support quality control

5) specification and control of the synthesis of plant machinery (i.e., ribosomes) that manufacture proteins.

Note #1: Amino acids do not combine spontaneously into proteins; proteins require a pool of the right amino acids, catalysts (which are other proteins), an input of energy, a recipe (i.e., DNA/RNA), and a manufacturing facility (ribosomes) in order to enable protein assembly.

DNA serves as the information storehouse for a finely choreographed manufacturing process in which the right amino acids (in the right isomeric form) are linked together with the right bonds in the right sequence to produce the right kind of proteins that fold in the right way to build biological systems.[37] More on this later.

Note #2: It is believed that DNA does not code directly for lipids or carbohydrates, but is involved with lipid and carbohydrate synthesis by coding for proteins (i.e., enzymes) that catalyze the reactions that synthesize the lipid and carbohydrate compounds needed by the cell. Most lipid synthesis occurs in the endoplasmic reticulum of a cell with the endoplasmic reticulum being a cell organelle consisting of a network of membranous tubules (see Fig. # 17).

Most DNA (in cells containing a nucleus) is located in the cell nucleus (where it is called nuclear DNA), but a small amount of DNA can also be found elsewhere in the cell (e.g., in mitochondria).

DNA is a long double helix molecule (like a circular staircase). For primitive cells such as single cell microorganisms known as prokaryotes, DNA exists as closed circular strands located in the cell's cytoplasm (since prokaryotes have no well-defined nucleus). Some forms of DNA in single cell microorganisms are known as plasmids.

For higher order cells such as cells in multi-cellular organisms, DNA is a larger molecule and is usually in a coiled linear configuration (i.e., not a closed circle) stored tightly in a package called a nucleosome, wound like a ball of yarn around histone proteins and a protein called chromatin.

> **Note:** If unwound into a straight chain, a nuclear DNA
> molecule would be about 6 feet long.

Since humans have several trillion cells, uncoiling and combining all the DNA in a human would result in a strand longer than taking several trips to the moon and back. In human cells, each DNA molecule (known as the human genome) contains 3 billion base pair molecules, organized into 46 chromosomes which, in turn, are organized into genes. Each base pair can be visualized as a rung on the double helix DNA "circular ladder like" molecule.

Most genes (i.e., those genes whose purpose has been determined) code for a specific kind of protein. Some genes are known to be involved in DNA repair mechanisms. By comparison to the size of DNA molecules in human cells, one of the simplest independent single cells ever discovered has 5 million base pairs in each DNA molecule which codes for over 4000 different proteins — still mind-boggling numbers regarding complexity.

The diagram on the following page shows a tiny portion of a DNA molecule.

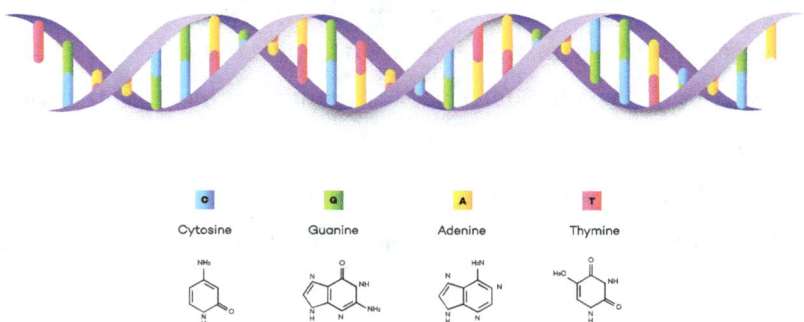

FIGURE 24: A SMALL SEGMENT OF DNA

Note #3: A Base Pair in DNA consists of two chemical molecules (i.e., bases) bonded to one another forming a "rung of the DNA double helix ladder." There are 4 different bases (i.e., molecules known as nucleotides) used in DNA:

- guanine (G),

- cytosine (C),

- adenine (A),

- thymine (T)

Each has a molecular weight of over 100. Adenine, e.g., is $C_5H_5N_5$ and has a molecular weight of 135.

There are a huge number (5 million to 3 billion) of chemical base pairs in a DNA molecule (depending on the type of cell or organism of which it is a part), and each base pair itself has a molecular weight of over 200 (so is not simple). So, a molecule of human DNA is like a tightly wrapped circular staircase, with 3 billion steps. That is, a molecule of DNA is extraordinarily complex (both in structure and function) although elegantly simple in that it is constructed entirely with just 4 different kinds of molecules.

Note #4: There is no particular chemical attraction of one type of base pair to another so there is no reason for base pairs to sequence on a DNA strand in any particular way based only on chemical attraction / bonding. Other directive forces must be in play.

A primary function of DNA is providing the blueprint (i.e., library of information or recipe) for making proteins. The genes, which are segments of DNA, involved for this purpose (about 20,000 to 25,000 in number for the human genome) only occupy 2-3% of the DNA in humans (and about 20% in bacteria). Some DNA is involved in the control of genes, with the content known as "regulatory DNA". That is, the instructions for turning genes on and off are written in DNA switches called regulatory DNA.

These switches are scattered throughout the non-gene regions of eukaryotic (e.g., human) cell DNA with some acting in concert with one another in response to signals received from the environment or elsewhere to control the activity (on, off, or expression rate) of particular genes. That is, regulatory DNA controls when, where, and how much genes are expressed.

Usually only a few of the many genes in DNA are active at any one time and the ones that are active have much to do with whether a cell is (or will be) a nerve, muscle, cartilage, or some other kind of cell. Some scientists used to refer to the large non-gene portions of DNA as "junk DNA", but, over time — scientists are prompting Mother Nature to slowly give up her secrets and the purpose and function of "junk DNA" is being discovered. One biologist has likened the gene portion of DNA as the notes of a symphony, with the regulatory DNA acting as the Conductor. So, as complex as DNA and cells have already been shown to be- much of the complexity has yet to be elucidated.

In starting the process of making a protein, the relevant gene in the relevant chromosome of DNA must be "activated" (meaning turned "on" — which probably involves regulatory DNA). The process then requires that the segment of DNA where the gene resides must be "unspooled" from the tight "ball of yarn" nucleosome (assuming eukaryotic cells), so that RNA can attach to the strand, pseudo copy it, make their way to cellular ribosomes, and start the process of making the desired protein based on the original DNA instructions (with DNA instructions akin to a recipe).

An unanswered question for naturalists/evolutionists is how the cell knows (when needing to make a protein) where to look for the appropriate gene among a cell's enormous collection of DNA and its 5 million to 3 billion ladder rungs. (Perhaps CRISPR, briefly discussed later, is an answer, at least for prokaryotic cells such as bacteria.)

> **Note #5:** DNA stores far more information in a smaller space than the most advanced supercomputer in the world. For example, one DNA molecule, which is microscopic, contains more information than 10 sets of Encyclopedia Britannica. Figure 25 illustrates the enormous gap between DNA and current man created storage devices in more detail. Note that the data density of DNA is many orders of magnitude greater than any man created

data storage device and that the energy required to store information in DNA is far less than storing the equivalent amount of information in any man-made device.

Which begs the question: if life started via natural phenomena, how did random acts of chemistry over time result in an information storage medium that is many orders of magnitude more efficient than anything that intelligent creative mankind has yet come up with.

	HARD DISK	FLASH MEMORY	DNA
Data Retention (years)	>10	>10	>100
Power Usage (watts per gigabyte)	.04	.01 –.04	10^{-10}
Data Density (bits per cubic cm.)	10^{15}	10^{16}	10^{19}
Data Capacity	Hundreds of Terabytes	Terabytes	Millions of Terabytes

1 bit = 1 memory location containing either a zero or one
1 byte = 8 bits = memory needed to store an alphanumeric character
1 terabyte = 1000 gigabytes = 1 million megabytes =1 billion bytes

FIGURE 25: THE STORAGE CAPACITY OF DNA VS. MAN MADE STORAGE DEVICES

Note #6: IBM's famous "artificial intelligence" (AI) computer Watson (which competed on the TV show Jeopardy in 2011), consumed about 10 refrigerators worth of space and consumed 85,000 watts of power. The far smaller human brain normally operates on about 20 watts of power.

Regarding RNA, cells do not use DNA directly in manufacturing proteins but, rather, a copy (using a very similar but not identical code) of the needed information from DNA is made onto shorter molecules called RNA (ribonucleic acid). If DNA is like a library book, RNA is like a scrap of paper with a key passage from the book scribbled onto it. DNA is often likened to a cookbook, with RNA being like a copy of a recipe.

RNA ends up in ribosomes (granular particles located in the cell's cytoplasm) which manufactures the proteins. So, ribosomes are like the chef or cook actually making the protein. RNA is similar to DNA, other than it is much shorter in length, uses a slightly different nucleic acid code, and only has one strand of nucleic acid building blocks (vs. the double stranded helix structure of DNA).

> **Note #7:** An average DNA molecule is more than a thousand times longer than an average RNA molecule.

As noted above, the process of converting the information in a RNA strand into a protein takes place on the surface of extremely elaborate and complex macromolecular machines inside the cell called "ribosomes". This process goes on in every living cell, even the simplest bacteria. It is as essential to life as eating and respiration.

There are many ribosomes in each cell, ranging from about 20,000 in *E. coli* bacteria to a few million in mammalian cells. There are over 50 enzymes (i.e., other proteins) in ribosomes that assist with the process of utilizing RNA information to manufacture proteins. Any explanation for the origin of life must show how this complex trinity — DNA, RNA and ribosomes — came into existence and started working together.

So, the process of using DNA (and different kinds of RNA) to make proteins is extremely intricate and complex. This is a problem for naturalists trying to explain the origin of life, because it is hard to imagine how something as complex as DNA, RNA, and ribosome molecular structure and function could ever have come about via naturally occurring events in the environment, with all three available at the same time and working together, within a protective cell semipermeable membrane (which itself contains proteins), as required for cell survival.

Another dilemma that natural phenomena scientists have yet to solve is: 1) proteins cannot be made without DNA / RNA and 2) cells cannot operate without proteins. It's the proverbial chicken and egg dilemma. Proteins and DNA/RNA each require the other— so which one came first?

There are a number of considerations in theorizing if a molecule of DNA could have been created by chance and/or natural selection. They include:

1. What triggers the unraveling of a tiny specific portion of the tightly wound DNA double helix to allow copying a specific DNA gene onto RNA (i.e., to start the process of making a specific protein)?

2. The earlier discussion regarding isomers of amino acids generally applies also to the nucleotides (i.e., bases) that make up DNA and RNA. A random mix of chemicals used to create nucleotide/ nucleic acids that make up DNA and RNA would presumably produce the different mirror image (i.e., left and right handed) isomers of these "building blocks." However, it turns out that strands of nucleic acids are made up only of right-handed nucleotide isomers. How can this be?

> **Note:** As with amino acids, isomers of a nucleotide molecule are identical regarding most characteristics and attributes. It has been difficult to determine how biological systems can separate one isomer of a molecule from another isomer of

the same kind of molecule, given how difficult it is to do in a scientist's laboratory.

Isomer purity must exist, as without it, the uniform "right handed" spiral of the DNA double helix would be a mess, curving this way and that.

As previously noted with amino acids, isomer purity is a characteristic of life. The mystery is how this selectivity emerged from the random prebiotic world.[38]

3. The best-known, man-created parallel to the construction of a DNA molecule is the commercial production of long chain polymers. DNA and RNA are polymers. Many plastics used in people's daily lives are polymers (e.g., PVC, polypropylene, polyester, polystyrene) used to make products such as rubber, nylon, epoxy, Teflon, piping and storage containers.

Famous physicist, Prof. Edgar Andrews, summarizing his own experience in working in industry with chemists making long-chain polymers, states: "It is impossible to build small molecules into long-chain polymers if there are impurities in the system, because the impurities 'poison' the chemical reactions and contaminate the catalysts that are needed to produce the polymers. It is ludicrous to suggest that amino-acids or nucleotides could spontaneously and consistently link together into long chains in a chemical environment containing a random assortment of many chemicals.[39]

In fact, the lack of certain chemicals (representing impurities) and energy sources used in experiments by scientists, which were likely present in primordial times, is a common criticism of many scientific experiments designed in hopes of supporting the natural phenomena origin of life. That is, many of the chemicals and/or energy sources (e.g., short wavelength UV light) representing primordial participants of competing or destructive reactions, are missing in such experiments.

4. In creating the first cell, not only would amino-acids have to link together to form protein chains (known as peptides), but, at the same time and place, nucleotides (the building blocks of DNA) would also have to link together into polymer chains, presumably in the presence of water (*note: cells are over 70% water by weight*). Such molecules could only be built from scratch if a variety of highly specific catalysts (i.e., enzymes — which are proteins) just happened to be conveniently on hand.[40] And how would such enzymes have come about, as the means to create them (i.e., DNA, RNA, and ribosomes) would not have existed yet?

    A potential concern in the life of a protein, as mentioned earlier, is the presence of water, which can cause hydrolysis reactions. Hydrolysis of proteins results in their being broken down into their component amino acids. Readers may wonder how proteins stay fully formed for long periods of time inside cells (which have an abundance of water) when they are readily broken apart in an animal's stomach (which also contains an abundance of water).

    Hydrolysis occurs in a stomach partially due to the presence of an acidic environment (pH 1-3, which causes proteins to quickly denature / unfold) and also to the presence of extracellular protease enzymes (e.g., pepsin) which promotes hydrolysis (the breakup of the protein). Proteins inside cells are also subject to hydrolysis, but the more neutral pH and the lack of enzymes specific to protein hydrolysis results in the reaction being extremely slow (i.e., a half life of many years), so is apparently negligible.

The potential concern with biomolecule degradation in suspected conditions near the earth's surface have led some scientists to believe that the origin of life did not occur on the earth's surface (e.g., near the surfaces of ponds or the oceans), but rather in hydrothermal systems at the sea floor and/or deep in the earth's crust which are devoid of sunlight, lightning strikes (for energy) or a supply of gaseous oxygen (see Fig. 13).

Note #8: Hydrothermal systems (i.e., vents) are openings in the sea floor out of which heated mineral-rich water flows. Conditions in such environments include temperatures up to 300 deg. C. and pressures up to 1000 atmospheres. Many of the properties of water (e.g., dielectric constant) are significantly different under such conditions. One of several issues with the hydrothermal theory is that amino acids (the building blocks of proteins) decompose rapidly in hot vents. Denaturing (i.e., unfolding) of any existing proteins (which occurs at elevated temperatures) would be another issue.

DNA and Cell Division

Another major function of DNA is to enable cell division in which DNA replicates itself and then is involved in creating two new daughter cells.

A primary concern of cell division is the maintenance of the original cell's genome (i.e., hereditary material coded as genes). Before cell division can occur, the genomic information that is stored in chromosomes must be replicated, and the duplicated genome must be separated cleanly between the newly forming daughter cells. A great deal of cellular infrastructure is involved in keeping genomic information consistent between generations.

Some cells (e.g., neurons and red blood cells) are not known to replicate themselves. Red blood cells are created by other means; i.e., bone marrow. Other types of cells, like skin, are constantly replicating. The human body replaces about 30,000 to 40,000 dead skin cells every minute. Considering all types of cells, the human body experiences about 10 quadrillion cell divisions in a lifetime.

Cells regulate their division by communicating with each other using chemical signals from special proteins called cyclins. These signals act like switches to tell cells when to start dividing and later when to stop dividing. It is important for cells to divide so as to, e.g., promote growth and enable cuts to heal. It is also important for cells to stop dividing at the right time. If a cell doesn't stop dividing when appropriate, cancer may eventually result.

The process by which a cell divides starts with a division of nuclear material (karyokinesis) which is then followed by a division of the cell body (cytokinesis), with each of the daughter cells receiving one of the two daughter nuclei.

The first step in DNA replication is to "unzip" the double helix structure (i.e., separate the two strands) of the DNA molecule. This is carried out by an enzyme called helicase which breaks the hydrogen bonds holding the complementary bases of DNA together (i.e., A with T, C with G).

Once "unzipped," other proteins are involved in creating duplicate copies of the now separate DNA strands. Then, other proteins are involved in creating the two daughter cells and collapsing the newly replicated DNA into its new nucleus home in its normal "ball of yarn" configuration. It is hard to imagine how such an unbelievably sophisticated and highly directed process could have evolved from basic chemistry and random natural events.

> **Note:** Regarding DNA, no man directed interaction of molecules has come close to producing from scratch this ultra-complex molecule and its embedded code which is essential to all known life. Man has been able to create new novel genes in the laboratory, but not a whole DNA molecule. They have used two different methods to create new genes:
>
> 1) cutting and pasting together segments of other genes
>
> 2) using highly specific chemical reactions and well thought out detailed procedures developed by brilliant scientists to link together the nucleotides A, T, G, and G in whatever order is desired.
>
> As of 2022, this second option is an expensive and time-consuming procedure and only practical to pursue for an individual gene (consisting of up to a few thousand nucleotides) — which is a tiny fragment ($< 0.01\%$) of an overall DNA molecule.

Further, a cell (with a couple of exceptions such as red blood cells that contain no DNA) can and must do something that no man-made factory can achieve, namely, automatically reproduce its entire self to order. It is this level of complexity that persuaded British evolutionist, B. Haldane, contrary to all his evolutionary convictions, to indicate that the first living organism could never have come into existence by chance. He stated that "we must give up the idea that such an organism could have been produced in the past, except by a similar pre-existing organism or by an agent, natural or supernatural, at least as intelligent as ourselves, and with a good deal more knowledge."[41]

r-DNA (recombinant DNA)

One of scientists' amazing creative accomplishments, beginning in the 1970s, has been their ability to modify DNA, resulting in "recombinant DNA" (r-DNA). The first commercial product making use of r-DNA technology was insulin (in a process developed at Genentech and manufactured and marketed by Eli Lilly & Co.).

In developing the process to make proinsulin (the precursor for insulin), the creativity of scientists resulted in their adding a gene to *E. coli* bacteria DNA which causes it to make proinsulin in addition to all the other proteins that the *E. coli* bacteria naturally makes. So much of the world's supply of insulin today begins via *E. coli* bacteria making insulin's precursor, proinsulin, in industrial bioreactors.

The process using r-DNA technology to make proinsulin was approved by the FDA and commercialized in 1982. A major benefit was that the insulin produced utilizing r-DNA is identical to human insulin, subsequently replacing previous commercial versions of insulin in which the insulin was extracted from bovine (cow) and porcine (pig) pancreas glands.

> **Note:** Animal insulin is similar but not identical to human insulin.

E. coli cell activity in making proinsulin still represents only a tiny portion of the overall functionality of man-modified *E. coli* cells. If adding this one function to *E. coli* cells required the creative intelligence of many brilliant scientists working for many years, is it not reasonable to believe that the remaining thousands of functions of *E. coli* (and other) cells also involved a creative intelligent force?

Original r-DNA techniques have since spawned the next generation of techniques that involve altering aspects of DNA, such as locating a particular gene within a DNA molecule and then performing certain functions, such as activating or deactivating the gene or cutting the DNA at the gene's location. One such method is known as CRISPR,[42] which involves a chemical system inside a cell that consists of two RNA snippets and an enzyme.

In its original form, one version of CRISPR acts like a homing device that guides molecular scissors (a separate enzyme) to a target section of DNA. Together, they (the two snippets and enzyme) work as a genetic-engineering cruise missile that disables or repairs a gene, or inserts something new where the scissor enzyme has made some cuts.[43]

While CRISPR is naturally occurring in prokaryotic cells such as bacteria, scientists are developing and using the method for additional purposes, including gene editing in eukaryotic cells. The technique is superior for gene editing in several respects compared to early r-DNA methods.

CRISPR and similar technologies have been patented and have given rise to several new biotech companies. Many clinical trials of potential commercial products, making use of gene editing technology, were underway as of 2020. The "gene editing" capability of this technology has already proven successful in the treatment of sickle cell anemia. But the question remains regarding CRISPR or any similar cellular system: How could such an amazing complex functional system have come about from a mixture of many chemicals and rolls of the dice?

Further details of the above topics are beyond the scope of this book.

A later section of this book will discuss the information content of DNA.

Cell Metabolism / Cellular Respiration

Metabolism is the totality of the large number of chemical reactions that take place within each cell of a living organism to sustain the cell (and pursue its specific mission if it is part of an organism). This is sometimes referred to as cellular maintenance. Metabolism involves the breakdown of incoming nutrients, providing energy for conducting vital cellular processes, synthesizing new organic material, and other reactions.

A perspective of the complexity of operations in a cell can be illustrated via a metabolic map of key chemical reactions occurring within a cell. A portion of such a map, common to most cells, is shown on the next page.

If the map looks like a highly detailed hard to read blur of complexity, that is intentional for the purposes of this book since this chart normally appears as a much larger, readable chart on the walls of bio-research laboratories. There are many thousands of different chemicals and chemical reactions within a cell and the above map shows only a portion. Compare this to typical chemistry courses (other than biochemistry) where examples and test questions usually only involve one or two reactions which are rarely more complicated than A + B <-> C + D. Non-biological industrial chemical manufacturing processes rarely utilize more than a few such relatively simple reactions while bio-processes utilize the complex metabolic machinery associated with living cells.

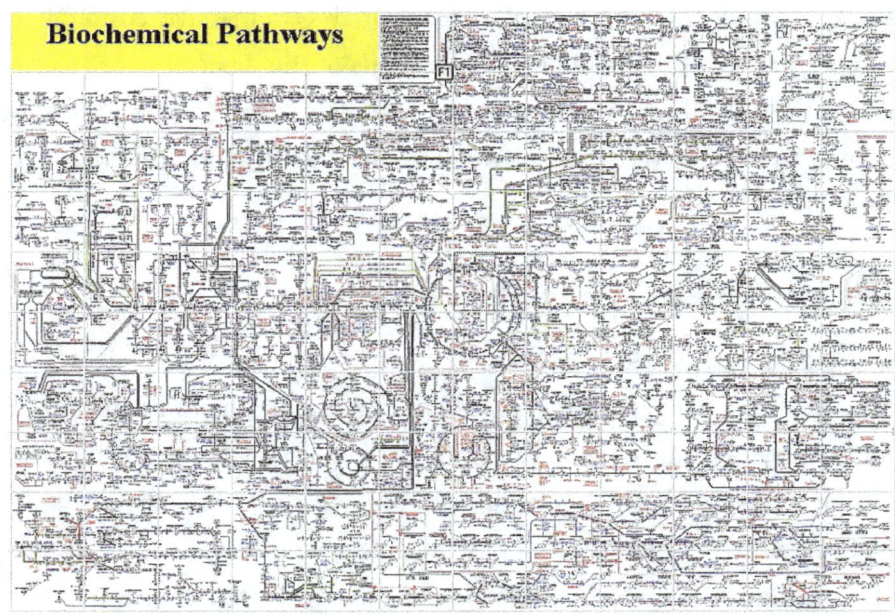

IMAGE CREDIT: AMAR, PATRICK &KEPES, FRANCOIS & NORRIS, VIC. (2012). PROCEEDINGS OF THE 2012 EVRY SPRING SCHOOL ON MODELLING COMPLEX BIOLOGICAL SYSTEMS IN THE CONTEXT OF GENOMICS.

FIGURE 26: A PORTION OF A LIVING CELL METABOLIC MAP

As biochemistry students know, each of the thousands of chemical reactions in the cell is governed by

1) an equilibrium constant (defining the ratio of products to reactants at equilibrium),

2) the concentration of reactants and products and

3) any relevant catalysts present (which help determine the rate of reaction in driving the reaction towards equilibrium).

So the rate of each of the many reactions in a cell are dependent on many factors.

Some of the chemical reactions shown in the metabolic map involve:

1) converting a main carbon source (often glucose) into energy- which is a process known as cellular respiration,

2) making cellular products (e.g., insulin by pancreas beta cells), and

3) the myriad of other functions involved in maintaining a living cell (or contributing to the organism of which the cell is a part).

Many of the compounds shown in the metabolic map are proteins, many of which are enzymes used to catalyze reactions. Readers are not asked to learn or remember the names of the chemicals in the metabolic map diagram or understand individual reactions; rather, the complexity of what is going on should be apparent. The cell is not a place where a few simple chemical reactions are occurring. Rather, it is a highly complex micro-factory of thousands of chemical reactions and chemical equilibriums, most utilizing complex molecules such as proteins, working together in harmony with cell organelles and/or transport mechanisms in the cell membranes, to accomplish the cell's metabolic functions.

The industrial chemical factories in the world, designed by highly educated scientists and engineers, and which use the most complex non-biological chemistry ever implemented by mankind, do not come anywhere close to approaching the complexity of chemistry inside the simplest of single living cells. In fact, rather than try and make certain products directly via traditional chemistry, scientists and engineers have deferred to the greater capability of microorganisms to help make some of society's most valuable products like penicillin, other antibiotics, insulin, growth hormone, vaccines, monoclonal antibodies, certain insecticides, bread, beer, yogurt, cheese, and biofuels.

For example, penicillin is made by a fungi (mold) microorganism. Most alcoholic beverages are produced through the fermentation of sugars (often glucose) by single celled microorganisms known as yeasts. That is, sugars are converted to alcohol (i.e., ethanol) and carbon dioxide by yeast cells. Most monoclonal antibodies are made using mammalian cells.

Fermentation is used to manufacture some bio-fuels (e.g., ethanol) in which microbes (yeasts or bacteria containing enzymes known as cellulases) is one option used to break down plant materials (e.g., the cellulose, lignin, and starches in agricultural waste products such as corn stalks, grasses, and wood chips) into simple sugars.

> **Note:** It is difficult to economically break down cellulose and lignin (which are key components of plant cell walls) via traditional chemical methods.

Microorganisms are then used to ferment the resulting simple sugars into ethanol, much like is done in making many alcoholic beverages. The ethanol produced is added, e.g., to petroleum fuels to help power cars. This mixture is cleaner burning than conventional gasoline, increases octane level, and creates fewer smog related exhaust compounds.

While there are many metabolic activities occurring inside cells, with cells sometimes referred to as chemical micro-machines, this book will briefly zoom in on one portion of the metabolic map as an example. The example is that of obtaining energy (from nutrients) for use by the cell. This topic is covered in almost all biochemistry and microbiology textbooks and is often the first topic taught in a biochemistry course.

To preface this section, this book up to here has used lightning and UV radiation as examples of energy sources that presumably drove primordial chemical reactions. That is, such sources must have existed in driving reactions in theorized primordial soups to enable small molecules to react to synthesize (i.e., create) more complex molecules. Of course, such energy sources can also be the cause of molecules breaking apart, such as UV radiation destroying RNA and DNA.

Anyway, occasional lightning strikes may be sufficient to enable a few small molecules to react to form larger molecules, but the energy capturing and management activities associated with supporting life are far more complicated than this. Usually, for established living organisms, transporting in nutrients from the external cell environment (rather than lightning strikes) is the usual incoming source of energy.

Living cells contain several energy-transferring molecules that enable the storage and efficient transfer of energy to power all of life's metabolic processes. ATP (adenosine triphosphate) is the final energy storage molecule in the energy storage pathway. Using a complex array of enzymes, cells harness the energy obtained from breaking down nutrient energy sources such as glucose. They then store that energy in the high-energy phosphate bonds in ATP.

The process is akin to charging a battery, though is far more complex. When stored "bond energy" in ATP is needed by the cell, the phosphate bonds in ATP are broken and the energy used as needed, with the ATP molecule then becoming adenosine diphosphate (ADP) or adenosine monophosphate (AMP). The ATP molecule is later recreated (similar to recharging a battery). ATP is the ultimate rechargeable battery. So, how could such a sophisticated rechargeable bio-battery have developed by chance?

Looking at the energy storage pathway in a little more detail, there is a small but critical portion of the metabolic map known as the Krebs Cycle, also known as the Citric Acid Cycle or the Tricarboxylic Acid (TCA) Cycle. Every cell utilizes the Krebs cycle — sometimes called the metabolic heart of life. It occurs in the mitochondria of cells. The primary purpose of the Krebs cycle is to capture energy from various molecules (e.g., originally from sugar nutrients) and store it (as ATP molecules) for use by the cells.

> **Note:** While all cells utilize the Krebs cycle, it occurs in different places depending on whether the cells are eukaryotic or prokaryotic.

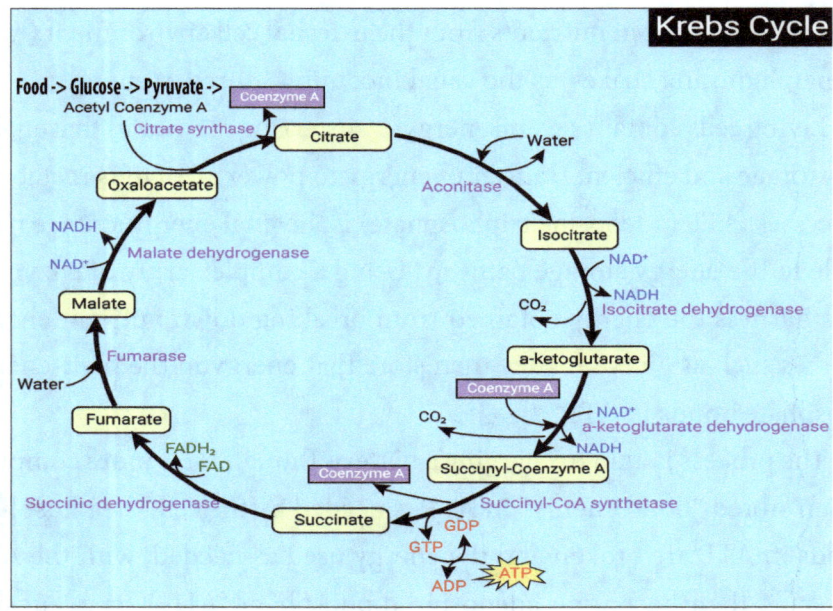

FIGURE 27: KREBS (TCA) CYCLE (CONVERTING NUTRIENTS TO STORED ENERGY)

The synthesis (i.e., manufacture) of all cellular materials can be traced to reactions originating within the Krebs cycle. In the oxidative environment that is standard with most living cells, this cycle also breaks down certain molecules (e.g., glucose) into smaller ones (i.e., it does not create larger molecules). So, the Krebs cycle operates to break certain molecules down, eventually to carbon dioxide and water, while converting energy in nutrients to stored energy in cells (i.e., ATP) which is then available for use in cellular activities.

> **Note:** in most plant (and algae) cells, photosynthesis is the process of transferring energy to ATP. In photo-

synthesis, the process begins when energy from light is absorbed by proteins. Then, many compounds and reactions ensue that result in energy stored as ATP which, in turn, is used to support various metabolic activities within the plant cells.

Some of what puzzles scientists (and one reason Fig. 27 is shown) is that the Krebs cycle itself is highly complex, requiring at least 10 different enzymes to function. How could these enzymes (which are proteins) have evolved and all be available at the same time and location, along with the components of the reactions they catalyze, if a functioning Krebs cycle is fundamental to life? It's another of the chicken-and-egg arguments that has stumped scientists.

The usual version of the Krebs cycle, shown in Fig. 27, assumes an oxidizing environment exists (which has been the case for at least the last tens of millions of years). Until recently, it was thought by some scientists that hundreds of millions of years ago, oxygen was not plentiful on earth, so a "reductive" (i.e., not oxidative) environment might have existed. Naturalists suggest that this might have allowed the citric acid cycle to operate in reverse, with the side benefit of creating larger biomolecules from smaller ones (i.e., from carbon dioxide and water).[44]

Many assumptions are involved with this theory. It so happens that a reverse Krebs cycle is known to be used by a few bacteria to synthesize carbon compounds. One question that arises is how a cell using a reverse Krebs cycle acquires and stores energy (i.e., the normal Krebs cycle generates energy, so one operating in reverse must apparently consume energy). So, how is ATP generated (needed for other cellular activities) in a reverse operating Krebs cycle?

The assumption of a more reducing atmosphere when life first occurred is probably invalid anyway, as by about 1990, atmospheric scientists established that Earth's atmosphere has been oxidizing (i.e., has had enough oxygen to create an oxidizing atmosphere) for the last 4 billion years — which includes the time that life is thought to have first existed.[45]

> **Note:** The conclusion of an oxidizing atmosphere during the past 4 billion years is hard to definitively prove so involves some assumptions.

The point is that theories raised by scientists to suggest certain possibilities, such as a reverse acting Krebs cycle, while well meaning, often raise more questions and issues than any light they might shed on a topic.

Finally, regarding metabolism, the cell's machinery (i.e., functionality) only operates correctly under a narrowly defined set of conditions. As industrial microbiologists know well, it takes a carefully concocted nutrient mix of carbon, oxygen, sulfur, and nitrogen sources, as well as a variety of minerals (with some needing to exist in a narrow range of concentration) in the cell's external environment to keep living cells happy and productive.

Also, temperature and pH ranges allowing cells to prosper are often narrow. For example, as mentioned earlier, high temperatures denature/ destroy proteins and so would kill their host cells. Many microbiologists have spent their career trying to optimize the nutrient soup and environmental conditions that are presented to and/or fed to microorganisms in order to sustain the cells and produce desired outcomes (e.g., commercial products).

These efforts are sometimes successful, but have required the intelligent creative work of scientists. Just presenting a random buffet of nutrients and environmental conditions to cells doesn't usually allow the cells to thrive and may even be insufficient to maintain life.

The Cell Membrane/Wall

Structure. Virtually all cells have a double-layered membrane made of phospholipids (i.e., lipid compounds containing phosphorus), other lipids (e.g., cholesterol) and proteins. (Fats are one form of lipid.) These molecules combine to form the cell membrane that contains and protects the inside of the cell, much like the outside walls of a house that contains and protects the inside rooms and furnishings. The nuclear envelope, which is a membrane surrounding a cell's nucleus, is also made up of phospholipids arranged in a lipid bilayer, as is the membrane of mitochondria (the parts of the cell that generate most of the chemical energy needed to power the cell's biochemical reactions).

A diagram of a small portion of a cell membrane is shown below:

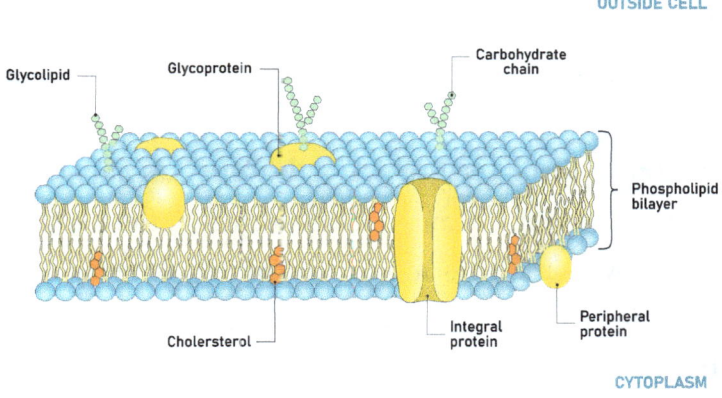

FIGURE 28: DIAGRAM OF A PORTION OF A CELL MEMBRANE

The internal workings of a living cell cannot function without these protective membranes. In addition to cell membranes, many (not all) kinds of cells have a cell wall adjacent to and outside the outer cell membrane. For example, plant cell walls are made out of cellulose. Fungal cell walls are made of chitin, the same material that insect skeletons are made of. Bacterial cell walls are made out of a mixed protein-sugar material unique to bacteria.

Regarding the cell membrane, a lipid bilayer membrane, on its own, would effectively seal the cell away from the external nutrients it needs and would also trap cellular waste inside the cell (like a house without doors or windows). That is, a living cell must be able to selectively transport certain molecules in and out of the cell. So, the membranes must also include a complex array of protein transporters to serve as cellular "doors." Because of their selectivity in what are allowed to come into and out of the cells, cell membranes are often referred to as semi-permeable membranes.

Lacking either the lipid bilayer or the protein transporters, a cell can't live. In essence, both the "doors" and the "walls" in a cell membrane/wall need to be present to sustain a cell and they had to have been present *from the beginning.*

Challenges in creating cell membranes. Many phospholipids, other lipids, and all proteins are complex molecules (with lipids usually being smaller molecules than proteins). It is unexplained as to how the appropriate mix of these many compounds appeared on the Earth and came together, even over time, to form a structured cell membrane/wall which permits the selective transport of the appropriate materials in and out of the cell to allow the cell to survive.

While it is now known how proteins are manufactured and replicated (via DNA and RNA), it is not well known how cells acquired the instructions to manufacture/replicate the lipid portions of cells (and there are thousands of different kinds of lipid molecules within cells). It is known that the cell's endoplasmic reticulum is involved in the synthesis of lipids.

It is also a mystery as to how cells acquired the ability to make cell membranes, especially when such membranes require combining molecules that aren't normally compatible from certain important perspectives. That is, lipids are hydrophobic (they dislike water) and most proteins are hydrophilic (they like water), so trying to create something that combines these two kinds of molecules is akin to trying to mix oil and water (which don't mix).

So, a question which naturalists have not yet explained is how a cell acquired the ability to manufacture and select certain highly specific enzymes (i.e., proteins that catalyze reactions) that then selected from a large number of different available lipids to combine with certain specific proteins (and other molecules) in non-naturally occurring reactions, to create cell membranes. That is, the blueprint to make proteins is known; i.e., DNA. So where is the blueprint that specifies even more complex structures such as cell membranes? Could it be located in the portion of DNA whose purpose is as yet unknown and which, for the time being, has been labeled as junk DNA?

Regardless, how did a primordial soup of thousands of different compounds come together in one place, required for cellular metabolism, and including the complex molecules (e.g., DNA/RNA) needed to manufacture proteins, such that a cell membrane was needed and was somehow created which "came into being" for containment purposes and to police the transport of certain molecules through itself? Where did the instructions (i.e., blueprints) come from that directed the construction of such membranes? Could this have all happened by chance?

Further, while cell membranes provide transport services for nutrients and waste products, there also needs to be available critical nutrients outside the cell to transport inside. For example, as was mentioned earlier, elements such as chlorine, calcium, sodium and magnesium are required for life. These have to be provided by an external environment since DNA and RNA do not create or replicate elements. So, where did all the 11 elements come from that are required for all cells, and so were required for the first cells on earth? Some are obvious (e.g., nitrogen), others not so.

Other Perspectives and Unanswered Questions Regarding Cells

Attempting to understand the basics of cell metabolism, reproduction, and cell membrane construction prompts a few additional considerations:

1. The total number of components in a cell to accomplish metabolism and reproduction exceeds the number of components used in some of man's most complex creations. For example, the Boeing 747 commercial plane has between 3 and 7 million parts (depending on the specific model of plane).

EVOLUTION OR INTELLIGENCE?

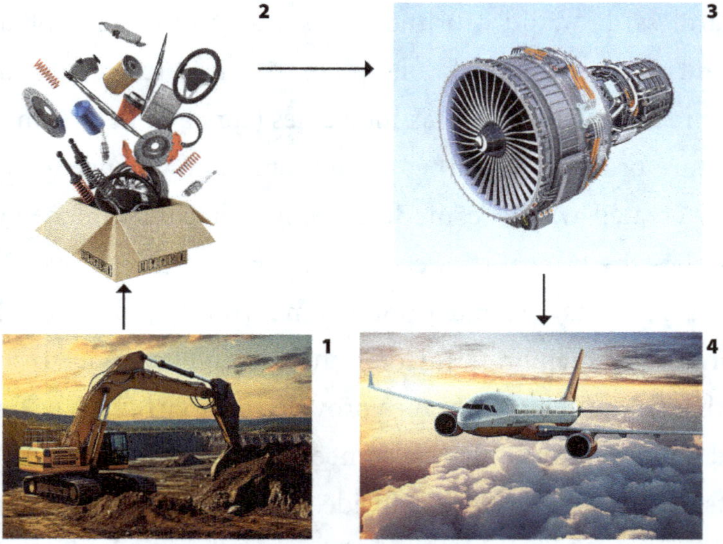

A 747 plane may have up to 7 million parts, while each living cell has millions more than this.

FIGURE 29: BUILDING A BOEING 747 AIRPLANE
Can one go from raw ores in the ground to a finished operational plane without directed intelligence?

The Empire State Building was constructed with about 10 million bricks. The number of components in a living cell is many times these numbers. For example, just the number of base pairs (i.e., rungs) in one human cell DNA double helix molecule is 3 billion. If the creative intelligence of man was needed to create such complex physical systems as airplanes and buildings, isn't it logical that some form of creativity was needed in putting together even far more complex structures and systems, such as a living cell?

CO2, METHANE, NH3, H2O+ENERGY

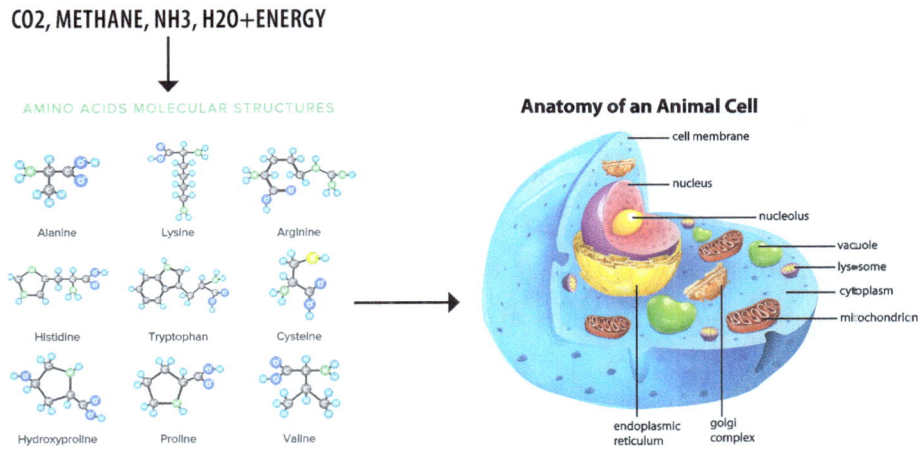

EVOLUTION OR INTELLIGENCE?

FIGURE 30: THE GAP BETWEEN SIMPLE MOLECULES AND A LIVING CELL.

Note: it would be more difficult to go from simple molecules that might have existed in the early earth's atmosphere (e.g., CO_2, N_2, methane, water) to a living cell via natural phenomena (i.e., without directed intelligence) than what would be required for natural phenomena (i.e., random events) to take ores from the ground and make a completed airplane (as illustrated in Figure 29).

Note: Astrophysicist Sir Fred Hoyle famously likened the chances of life arising spontaneously (from existing molecular precursors) to the probability that a whirlwind in a junk yard (or even a Boeing components warehouse) could assemble a Boeing 747 aircraft.[46]

It isn't nearly enough to get all the ingredients together in one place and expect spontaneous generation to occur. And this applies to even the simplest system within the cell, let alone the whole living cell. **That is, the creation of a living cell (or a building or a Boeing 747) requires ingredients <u>plus information/instructions plus organization plus creativity</u>.**

2.   The evidence from studying DNA mutations (which is the basis of evolution) points to a greater likelihood of genome degradation over time rather than upward evolution. I.e., the large majority of mutations are detrimental to the cell. No known "beneficial" mutation does anything but produce a new strain or variety within the same species, and the changes are trivial in the sense that no new biological structures are ever formed.[47]

3.   Scientists do not agree as to what aspect of cells came first if the natural phenomena hypothesis is correct: — DNA, RNA, a mix of proteins, a cell membrane/wall, or some other part of a cell. Any arguments would appear to be academic as most or all of the above-mentioned components would have had to exist at the same time at the same place (like parts of a watch) in order for a cell to exist and live.

However, there is still significant discussion in the literature suggesting that DNA and RNA might not have originally coexisted- with perhaps RNA preceding DNA. That is, some natural phenomena scientists, at one time, apparently believed that the earliest cells may have utilized RNA to contain cellular "blue-print" information with the more complicated DNA coming along later. RNA is a simpler molecule (similar to a small piece of DNA) that is more functional than DNA. I.e., RNA contains portions that mimic part of the genetic code and, also, is actually part of the mechanism required to manufacture cellular proteins, a requirement for life.

Also, for a single strand of RNA to replicate (which is required if RNA was the forerunner of DNA), there must be an identical RNA molecule close by. To have a reasonable chance of having two identical RNA molecules of the right length would require a library of many trillions of RNA molecules, and that effectively rules out any chance origin of a primitive replicating system driven only by RNA.[43]

If RNA preceded DNA, there is still no hard evidence that RNA can do all the things that natural phenomena or evolutionary theory would demand of it. It is a valuable molecule (e.g., acting as a transitional molecule between DNA and protein synthesis, being an on-off switch for some genes, etc.), but it is not functional enough. For example, if life began with an RNA molecule (without DNA), then that RNA molecule must have been able to make copies of itself; i.e., it should have been self-replicating. But no known RNA can self-replicate. Nor can DNA. It takes a battalion of enzymes (i.e., proteins) and other molecules to build a replica copy of a piece of RNA or DNA.

Left unexplained is how RNA itself first arose. Like DNA, RNA is a complex molecule made of thousands of smaller "building block" molecules called nucleotides that link together in very specific, patterned ways. While there are scientists who think RNA could have arisen spontaneously on early Earth, others say the odds against such a thing happening are astronomical.

"The appearance of such a molecule (i.e., RNA), given the way chemistry functions, is incredibly improbable. It would be a once-in-a-universe long shot," said Robert Shapiro, a chemist at New York University. "To adopt this view, you have to believe we were incredibly lucky."[49]

No plausible mechanism has been demonstrated by scientists to make all three components of an RNA building block — the sugar ribose, a phosphate group, and a base, and then link them all together in just the right way (i.e., string them end-to-end) to make an RNA strand. That is, no one has figured out a way to synthesize RNA from the molecules in the "Primordial Soup". No known mineral acts as a template (i.e., catalyst) to assemble sugar, phosphate, and base into a nucleotide, much less link nucleotides together into a strand of RNA.[50]

Anyway, for some of the above listed issues, the theory that RNA preceded the arrival of DNA has subsequently been largely discounted by the scientific community.

4. A complex biological system, (e.g., cell, organ) needs to conform to the principle of irreducible complexity.[51, 52, 53] A system is irreducibly complex if it has a number of different components (or subsystems) that all work together to accomplish the mission of the system. A clock is an example of such a system, as it will not show the correct time of day unless all its internal gears and other moving parts are present, work-

ing properly, and driven by a usable available source of energy. A living cell is a great example whereby many systems must all be existing and working together for the cell to function. If one of the systems (e.g., cell membrane or ribosomes) were to be removed, the cell would no longer function. For organisms, the eye and ear are examples of irreducibly complex subsystems. An irreducibly complex system is highly unlikely to be built piece-by-piece through natural phenomena processes, because the system has to be fully present in order for it to function.

As an analogy, consider a Boeing 747 airplane which consists of up to about 6 million parts, none of which can fly on their own. So, just as the key parts of an airplane need to be assembled before the plane can fly, so the parts and systems of a living cell need to be in place or it can't function. There are literally thousands of such "subsystems" in a single cell that are vital for it to operate. One of the thousands of such subsystems, described previously, is the Krebs cycle which is needed to obtain and store energy as ATP. Another example (in certain cells) is the one that senses light and turns it into electrical impulses for further use by the cell.[54]

Evolution can't produce an irreducibly complex biological machine suddenly, all at once, because it's much too complicated. And it can't produce it directly by numerous, successive, slight modifications of a precursor system, because any precursor system would be missing a part and consequently couldn't function. There would be no reason for it to exist. And natural selection only chooses systems that are already working.

Michael Behe, Professor of Biological Sciences at Lehigh U. and a Senior Fellow at Discovery Institute's Center of Science, believes that irreducibly complex systems are strong evidence of a purposeful, intentional design by an intelligent agent. No other theory succeeds, certainly not Natural Phenomena.[55]

Currently, there is only one principle that is known that can develop complex interactive systems, and that's intelligence. The only force known to be able to make such irreducibly complex systems (e.g., machines) is intelligent design.[56]

CHAPTER 7

A Library of Information: Taking a Closer Look at DNA

> **Objective:** Explore the information content of DNA noting that DNA is far more than a traditional chemical/biological structure associated with a set of chemical reactions. View DNA much like a cookbook, containing recipes as to how to make things.

Earlier in this book, the structure and basic functions of DNA were reviewed. What follows in this section is a closer look at the language that DNA utilizes and the information that DNA contains. Molecules such as nucleic acids (which make up DNA and RNA) are very unusual compared to most other molecules; i.e., they (as parts of sets) are informational molecules. They possess the characteristic of storing and transmitting information.[57]

> **Note:** Information is neither matter nor energy, so is something different than what is covered by the laws of chemistry and physics that describe material and energy phenomena (e.g., conservation of mass and energy). That is, because information is not matter, its origin cannot be explained by material processes such as chemical reactions.

DNA is like a library. An organism accesses the information that it needs from DNA so that it can build, sustain (and even self-repair) many of its critical components. That raises the question of the origin of the information. If one can't explain where the information comes from, then life has not been explained, because it's the "information" that describes how molecules such as proteins are structured and integrate into something that actually functions.

Human DNA (in each cell) contains more organized information than the Encyclopedia Britannica. If the full text of the encyclopedia (or something equivalent) were to arrive in decipherable code from outer space, most people would regard this as proof of the existence of extra-terrestrial intelligence. But when seen in nature on earth (i.e., with DNA being a much larger structured source of information than the Encyclopedia Britannica), it is explained by some people as the workings of random forces.[58]

During the past several decades, several radio telescopes, some operating independently and some networked together into arrays, have been built that monitor the universe for electronic radiation in the radio signal portions of the electromagnetic spectrum. Electronic signals are received and then analyzed as to whether they represent random noise or if they contain any decipherable information. NASA sponsors some of these programs, e.g., SETI (Search for Extra-Terrestrial Intelligence). Intelligible communications in these signals would be widely hailed as evidence of an intelligent source. The question is — why then doesn't DNA's message sequence also constitute evidence for an intelligent source? After all, DNA information is not just analogous to a message sequence such as Morse code; it is this type of message sequence.[59]

FIGURE 31: A RADIO TELESCOPE USED FOR NASA'S SETI PROGRAMS

The four base molecules (i.e., nucleotides) in DNA (abbreviated as G, C, A, and T), which make up the 3 billion base-pair "letters" (i.e., double helix ladder rungs) in the human genome, constitute the four-symbol alphabet of the "language of life." That is, DNA stores information in the form of a four-character digital code (much like computers store information in the form of a two-character digital code, i.e., zero and one).

The bases in DNA, made from the 4 different kinds of letters, are first arranged into word-like triplets called "codons" with each codon (consisting of 3 bases) corresponding to (i.e., coding for) a specific amino-acid (of which there are 20 in the standard genetic code — plus two more amino acids available for special purposes. A codon is like a 3-letter word in the English language. Since each of the 3 letters in a codon can have any of 4 bases, the number of different items that a codon can specify is $4^3 = 64$.

So, 20 (or 22) of these 64 items correspond to individual amino acids; others are, e.g., punctuation functions which indicate the beginning or ending of a gene sequence. For example, if a codon in a gene in DNA contains the sequence GCA, this corresponds to the amino acid alanine which will then occur in the resulting protein being manufactured. So, alanine will occur in the same place in the sequence of amino acids in the manufactured protein as its place in the gene sequence.

The sequence of codons (i.e., 3-letter words) that specify the sequence of amino acids that make up a specific protein is called a gene. The use of codons is the first level of DNA organization and without it, life could not exist. So, if there were no base-pair triplets (i.e., codons), there would be no way of knowing which amino-acids to select to make proteins. No proteins, no life. The way specific codons specify (i.e., code for) particular amino-acids is called the genetic code.[60]

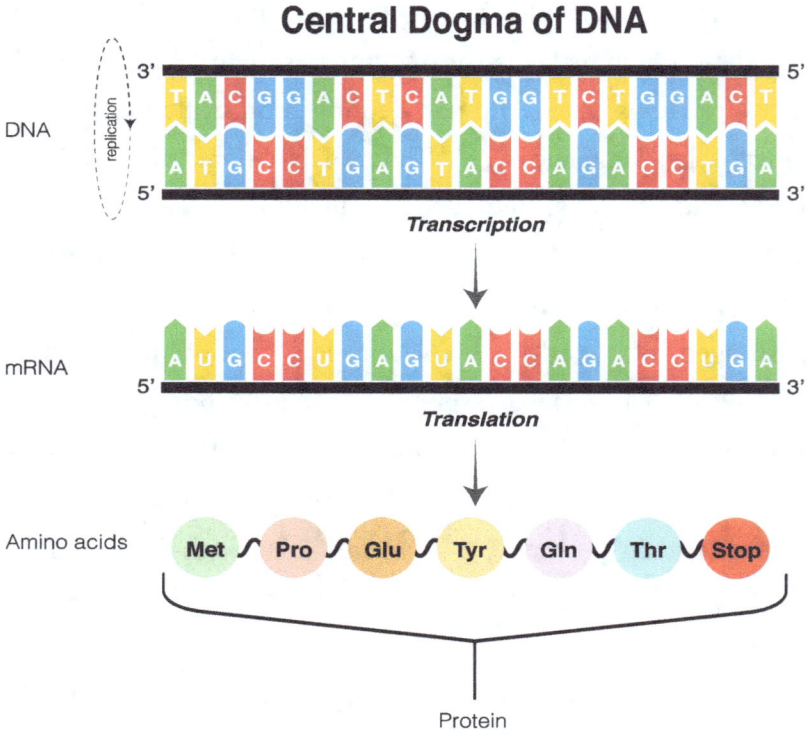

Central Dogma of DNA

FIGURE 32: PRODUCING A PROTEIN FROM DNA
The first 3 nucleic acids shown, A-T-G, is a codon, which becomes A-U-G in RNA, which is the code for the amino acid methionine (Met).

DNA (a nucleic acid) is far more than a chemical sequence of nucleotides that may somehow have come together from unexplained sources. It is also a complex data and knowledge base of information, utilizing an alphabet of 4 characters (the 4 base units) — which contains the detailed instructions (genes) for making proteins and which is complete with all the appropriate "punctuation" characters that indicate, e.g., when a gene sequence in the DNA strand starts and stops.

It took intelligent man to create alphabets used elsewhere in communication systems to transfer information (e.g., the dots and dashes of Morse Code, 0-1 bits in computers, the 26 letters of the English language). So, how did the alphabet in DNA come about and how did the cell learn to utilize it to accomplish all the amazing functions that DNA enables? It would seem some form of intelligent design must have been involved.

That is, how did cells learn how to translate DNA information (with its 4-character nucleotide alphabet) into the code for proteins (with its 20 amino acid alphabet)? This means converting one structured language into another.

So, even if multiple highly improbable activities occurred to create DNA and protein molecules by some evolutionary and natural selection process, a serious problem remains. Since the independent formation of proteins and DNA would have supposedly occurred by random chemical reactions, the sequence of amino acids in the proteins and of base pairs in the DNA would bear no relation to each other. There would be no match between DNA codons and the amino-acids for which they need to code. That is, the dictionary (analogous to the Rosetta Stone) required to translate base-pair sequences into amino-acid sequences would be missing.

If such translation is not available, life never gets started — i.e., "it is lost in translation."[61]

> **Note:** It is not the chemistry of DNA that underwrites life but the information stored in the DNA — information that is encoded by the meaningful sequence of base pairs and which spells out instructions that the living cell can read, translate, and use. Life, therefore, consists not only of molecular chemistry but in the information stored by the molecular chemistry — which is altogether different.

So, where did the information come from in creating DNA and the translator?

If a chemist were to make a DNA molecule identical to that in common *E. coli* bacteria, about 4,600,000 instructions would be needed to specify the sequence of the 4 kinds of letters in the DNA molecule.[62] In creating actual life (i.e., the living *E. coli* cell), where would those instructions have come from?

> **Note:** the above is a hypothetical. No chemist has ever made a DNA molecule from non-biological matter.

The storage and use of information in the living cell exhibits many, if not all, of the characteristics of human language — summarized as follows:[63]

A. As in human language, the cell employs a DNA code, specifically a four symbol (i.e., molecule) alphabet.

B. As in human language, the cell organizes its symbols into words (i.e., codons).

C. As in human language, the words have an agreed meaning so that they can be recognized by the ribosomes and "translated" into an alternative amino-acid language.

D. As in many human languages, punctuation is used to demarcate (i.e., separate) genes

E. As in human language, words are organized into instructions (e.g., specifying which one of many possible proteins is to be made by copying a given gene).

F. As in human language, the cell's language has a purpose — namely to construct protein sequences that will, in turn, fold in specific ways to provide functional capability and catalysts needed for the cell's operations.

In other words, the molecular information system in living cells not only resembles a language — it is a language — and it all looks remarkably like intelligent design.

The "Universal Definition of Information" indicates that it contains four traits: Code, Meaning, Expected Action, and Intended Purpose.[64] DNA fits this description (i.e., checks all the boxes) in that it:

1. Contains code; i.e., the 4 nucleotide letters A, T, C, and G which make up 3-letter words (codons), arranged linearly in a sequence within DNA

2. Meaning: each codon represents one of 20 specific amino acids to use in constructing a protein

3. Expected action: instruct the building of proteins (plus probably many other functions not yet understood)

4. Intended purpose: the existence of life.

The First Law of Information states that "Information cannot originate in statistical processes. Chance plus time (which represents a statistical process) cannot create information no matter how many chances or how much time is available." Recall, information is neither matter or energy.

The Second Law of Information states that "Information can only originate from an intelligent sender."[65]

"DNA is an information code...The overwhelming conclusion is that information does not and cannot arise spontaneously by mechanistic processes. Intelligence is a necessity in the origin of any informational code, including the genetic code, no matter how much time is given."[66]

So, information is stored in DNA molecules, transcribed onto RNA molecules and translated into proteins. The process of storage, transcription and translation closely mimics an advanced human language, involving codes, syntax and semantics. This is the "language of God" (ref. Francis Collins, leader of the Human Genome Project). It follows that the essence of life resides not only in chemistry but in information and communication: i.e., things that can only be the product of intelligence, not chance, and which the hypothesis of a creator leads us to expect.[67]

In DNA, each individual base, (i.e., letter of the DNA alphabet), is chemically bonded to the sugar-phosphate backbones (i.e., side rails) of the DNA molecule. That is how they're attached in the DNA's structure. But, — and here's the point — in creating a DNA double helix molecule, there is nothing chemically that forces base pairs into any particular sequence. The sequencing has to come from somewhere else — which suggests some form of intelligence. That is, in DNA, neither chemistry nor physics naturally arranges the letters (i.e., A, T, C and G) into the assembly instructions for proteins. Clearly, the cause comes from outside the system. And that cause is intelligence.[68]

Scientists are taught to assess each hypothesis on the basis of its ability to explain the evidence at hand. Typically, the key criterion is whether the explanation has "causal power," which is the ability to produce the effect in question. In this case, the effect in question is information. Neither chance, nor chance combined with natural selection, nor self-organizational processes have the causal power to produce information. But there is one entity that does have the required causal powers to produce information, and that's intelligence.[69]

The coding regions of DNA have the same relevant properties as a computer code or language. Whenever one finds a sequential arrangement that is complex and corresponds to an independent pattern or functional requirement, this kind of information is always the product of intelligence. Books, computer codes, and DNA all have these properties. We know books (using an alphabet) and computer codes are designed by intelligence and so the presence of coded information in DNA also implies an intelligent source.[70]

Natural phenomena cannot answer the fundamental problem of how to get from matter and energy to biological function without the infusion of information from an intelligence. Naturalistic theories that rely solely on matter and energy are not going to be able to account for information. Only intelligence can.[71]

It was fitting that when scientists announced that they had finally mapped the three billion "letters" of the human genome — a project that filled the equivalent of 75,490 pages of the New York Times newspaper, President Bill Clinton said scientists were "learning the language in which God created life," while geneticist Francis Collins, Head of the

Human Genome Project, said DNA was "our own instruction book, previously known only to God."[72]

Geophysicist Stephen Meyer, Ph.D., has stated: The origin of information in DNA — which is necessary for life to begin — is best explained by an intelligent cause rather than any of the types of naturalistic causes that scientists typically use to explain biological phenomena.[73]

Einstein may have summed up this topic best when he said: "God does not play dice." He was right, GOD PLAYS SCRABBLE.[74]

CHAPTER 8

The Law of System Order/Disorder

Objective: Compare the tendency for most objects and systems to naturally decay over time (i.e., become more disordered) to the construction of the highly complex living cell. Ponder if a cell's construction (with its thousands of components and multiple subsystems all working together) could have occurred via natural phenomena random events.

There are certain universally accepted laws of chemistry and physics that are associated with the world that we live in. Two of these are the well know "conservation of mass" and "conservation of energy" laws.

While perhaps less well known, another of the laws is sometimes referred to as the 2nd Law of Thermodynamics (or Law of Entropy). In layman's terms, this is the law of physics which indicates that systems spontaneously evolve towards states of higher randomness and disorder (often referred to as entropy). That is, systems don't naturally become more structured and ordered; rather, the opposite is true. Without external intervention, they eventually decay and fall apart.

101

FIGURE 33: A RUSTING CAR
(INCREASING ITS LEVEL OF DISORDER; I.E., ENTROPY, OVER TIME)

One of the science principles involving entropy is that the statistical probability that a given system (or arrangement) will arise spontaneously in nature is inversely related mathematically to the degree of order or complexity of the system. Systems with low complexity (high entropy) are more likely to arise spontaneously while systems with high complexity (low entropy) are unlikely to do so.[75] Nothing that creative mankind has created is anywhere near as complex as a living cell — so it is statistically unlikely that the first living cell(s) arose spontaneously.

As examples of non-biological systems, consider the state of an existing automobile or building or computer system. If left on their own, they will eventually fall into a state of disrepair (i.e., higher disorder) and fall apart. For example, mechanical and electrical systems in buildings, if left on their own, subject to the natural forces of nature, without intervention by man, will never get updated, become more reliable, have electrical circuits added, grow bigger, or repair themselves. They will never evolve on their own to become a more functional, modern, weather resistant, or smarter system. Rather, pipes will rust, corrode, and start leaking, pumps will wear out, batteries will lose their strength, supply voltage surges will weaken electrical components, mold might set in, and weather conditions will slowly degrade the system.

If waiting long enough, such systems can become so disordered that they become junk (or dust). Would any Ford Model T automobile be running today if not for the external involvement and tender loving care of intelligent owners and antique dealers who changed the oil, replaced parts, added protective coatings to surfaces, and stored the car in weather protected environments? The weather is even taking its toll on such celebrated structures as the pyramids, Roman Coliseum, and Greek Parthenon.

Existing biological systems are similar to degrading non-biological systems in many respects. No organism lives forever. As they age, and despite the efforts of brilliant doctors and medical researchers, organisms become subject to an increasing range of maladies, with one or more resulting in the organism's eventual death. That is, the highly structured order of living tissues will eventually succumb to a state of disorder due to, e.g., DNA mutations within cells, tumors developing, natural cell death (known as apoptosis, necrosis), injury, and/or disease. As is noted in many funerals, our bodies will eventually return to "dust" (as the ultimate form of structural "disorder").

Regarding DNA, mutations frequently occur, manifested as a change within a DNA sequence. Mutations can result from DNA copying mistakes made during cell division, exposure to ionizing radiation, exposure to chemicals called mutagens, or infection by viruses. When a gene mutation occurs, the nucleotide building blocks are in an incorrect order (or have some other flaw) which means the coded instructions are incorrect and faulty proteins are made or chemical control switches (that turn certain chemical reactions on and off) are changed. The huge majority of mutations are detrimental to the cell; i.e., the cell can't function as it is supposed to.

A mutation that improves the capability or performance of a cell is very rare. However, mutation is, ultimately, the only way in which new variation enters a species.

As additional perspective, the up to 2 million protein molecules in a cell are able to function, in part, by virtue of their 3-dimensional structure, which includes how the protein molecules are folded. A protein is only able to maintain its 3-dimensional folded structure within a defined range of environmental conditions, one of which is temperature. That is, if temperature becomes too high, protein molecules unfold (known as denaturing) and the cell dies.

This is the basis of one of the main kinds of sterilization operations used in industry (i.e., heat sterilization). Heat sterilization is the process of taking matter (contained, e.g., in bioreactors or autoclaves) up to a defined temperature (e.g., 121 deg. C.) for a specific amount of time (e.g., 20 minutes) which kills all living organisms (including fungal spores).

Actually, if spores are excluded, subjecting a microorganism to 60 deg. C. for 5 minutes is more than sufficient to kill it, — with 60 deg. C. used for many industrial cell harvesting operations requiring microorganism kill. This (i.e., protein denaturing) is just one of many processes or conditions that contribute to an increase in the "disorder" of a biological system.

Radiation is another of these processes, as, e.g., UV radiation can kill cells directly as well as cause massive mutations in cellular DNA, causing eventual cell death.

Body cells are constantly dying. About 10 billion cells in a human body die each day. The rate of cell regeneration (off-setting some of the cell death) eventually slows down and the body dies.

An average human cell can replicate itself about 40 to 60 times before the DNA or other parts of the cell become so damaged that the cell can no longer function. This damage is associated with apoptosis with the limit known as the Hayflick Limit.[76] This means that there is a programmed-cell-death and natural age limit to cells and, thus, the organisms of which they are a part.

> **Note:** Brain cells (of which there are about 100 billion) have less ability than other kinds of cells to regenerate and, with few exceptions, cannot regrow after damage.

So, if most things (biological and physical) in the world are tending toward an ultimate state of disorder, how is it that biological organisms actually grow and become more "ordered" during at least the early part of their life cycle? That is, babies grow, learn, and become increasingly more capable and productive as they become adults.

This is actually a testament to the complexity of living organisms and the force(s) that created them. That is, no man has ever created anything that, once created and left on its own, grew bigger and increased its range of functions. How is it that biological systems, or even individual cells, can go against the natural forces driving systems/cells towards "disorder" and can do things that even the world's most capable scientists cannot create in the laboratory or construct the physical equivalent of? Could a more intelligent creator be involved?

Some scientists historically have wondered if the primary growth period of an organism (i.e., birth to adulthood) represents a violation of the 2nd Law of Thermodynamics (since, e.g., animals become more ordered as they grow in size and complexity during the first part of their life cycle). In response, other scientists have pointed out that the 2nd Law of Thermodynamics only applies to closed (i.e., contained) systems and not to "open" systems for which external energy sources (such as food and UV radiation) can affect a system and its degree of order, sometimes even increasing a system's order. That is, while the natural tendency in nature is to dissipate energy through space, that certain phenomena in nature (e.g., lightning strikes) are capable of concentrating energy — which would be required to create the molecules and structures necessary for life.

So, while most activities inside living cells occur within cell membranes, which are characteristic of closed systems, the existence of cell membrane / cell wall portals to the outside world and the requirement for energy containing nutrients to be provided from the outside world argues as to cells in nature being open systems and, therefore, not governed by the second law of thermodynamics.

A simplistic analogy exists in making a balloon grow bigger. It only happens when an external supply of pressurized gas (i.e., energy source) is provided and a properly shaped elastic well engineered membrane of the proper thickness and strength exists. Otherwise, the balloon will not inflate, or if already inflated and the energy source is unavailable, will eventually lose its shape and dissipate. Even if the external energy source (i.e., pressurized gas) continues to be available, various environmental events (e.g., pin pricks) can cause an inflated balloon to dissipate.

It is known that the important macromolecules of living systems (e.g., proteins, DNA) are more energy rich than their precursors (e.g., amino acids, sugars). Hence, classical thermodynamics predicts that such macromolecules will not spontaneously form. Therefore, external energy sources are required to allow macromolecules to form, as well as mechanisms to convert such energy into useful work.[77]

So the development of cells and complex living organisms is not thought to have violated any law of thermodynamics, but only if external sources of energy were available (which apparently there were via, e.g., UV light, lightning strikes, nutrients) when life began and if mechanisms existed for, e.g., creating higher energy macromolecules (e.g., proteins, RNA) from lower energy building blocks such as amino acids.

The identification of such "origin of life" mechanisms is troubling since the current understanding of such mechanisms requires that proteins (i.e., enzymes functioning as catalysts) existed when life began and it is not known how such enzymes could have existed in the primordial soups presumably leading to the origin of life. L. Lester and R. Bohlin's book *The Natural Limits to Biological Change* makes an analogy to an automobile where there is plenty of external energy available from filling the gas tank and then combusting the gasoline, but useful work is only accomplished by virtue of a highly structured pathway characterized by an engine, transmission, drive shaft, universal joint, and wheels (which, incidentally, required human's intelligence to design and implement).

Another example is that of photosynthesis in plants. The sun is a source of energy for plants, but that, by itself, does not permit plants to remain alive and grow (e.g., take water and carbon dioxide and manufacture sugars and oxygen). Complex metabolic photosynthesis metabolic machinery, including several enzymes (proteins), exists to convert sunlight, water, and carbon dioxide into useful chemical energy for the cell's use.

It is hard to imagine how such complex machinery could have already been in place via natural phenomena when the first plant cell came into being. So, the structured pathways that are now known to be required for macromolecule synthesis (including use of DNA/RNA information codes) could not have evolved from natural processes when life first arose.

In an analogous way, UV radiation from the sun (an available energy source since the beginning of life on earth) is not directly beneficial by itself to human life. In fact, UV radiation is usually detrimental, especially in excessive amounts, as it can cause sunburn, DNA mutations and skin cancer. But in moderate amounts and in concert with a biological mechanism to make constructive use of certain UV wavelengths, it can catalyze chemical reactions in the skin involving cholesterol and other compounds, leading to the creation of Vitamin D which is beneficial in facilitating absorption of calcium from the intestines, which, in turn, strengthens the skeletal system.

So again, increasing the order (complexity) in a system requires energy and appropriate mechanisms for utilizing the energy. The sun can supply the energy in some cases, but a key question is how did a cell's energy conversion and utilization mechanisms come into being? As described earlier in this book, a similar analogy exists with nutrient chemical energy (e.g., with glucose) where a complex cell metabolic system (i.e., the Krebs cycle) exists to convert such energy into energy the cell can utilize (i.e., ATP).

In summary, all systems that man has created trend towards natural destruction (increased disorder) when left on their own due to the natural forces of nature. Further, all individual biological entities also eventually trend towards disorder and die. To offset this tendency towards disorder, especially for the early portion of life cycles in biological systems, there must have been some statistically highly unlikely random and exceedingly complex chemical evolution process taking place (unexplained so far by scientists), requiring the existence of external energy sources and structured metabolic pathways allowing for captured external energy to result in useful work — or else, some intelligent creative process occurred that intelligent man has not yet been able to mimic in anything he has created.

CHAPTER 9

Statistical Probabilities

Objective: Cite some analogies and published calculations that indicate the likelihood (or lack thereof) of natural phenomena as the cause for the origin of life having a reasonable chance of being correct. Compare the probability results with the definition of the statistical equivalent of zero.

Many articles in the literature cite statistical probabilities as to the likelihood of various cellular components or systems in living cells developing from random or natural phenomena causes. Probability calculations are always based on various assumptions. Many of the published statistical probability calculations have been questioned by one natural phenomena advocate or another as inaccurate, irrelevant, and/or using invalid assumptions.

However, most scientists believing in the natural phenomenon theory, while agreeing that the probabilities involved are extremely small, seem to have "hung their hat" on the immensity of the universe with its billions of stars and planets and the millions of years involved that arguably allow for even the most unlikely of events to have occurred (with Earth being the lucky winner or one of very few winners).

To help introduce the content in this chapter, note that in recorded history, no known recorded flip of 100 coins has ever resulted in all heads (or all tails), but it is still theoretically possible to happen — which encourages natural phenomena proponents to keep believing natural phenomena resulted in life, despite its exceedingly low probability (almost zero) of being true. The main example in this chapter will show that the probability of even a single protein being formed by random chance is far less than getting 100 heads (or 100 tails) in a row in a coin flip.

The conclusions noted in this chapter (and book) are not dependent on the accuracy of any one or two of the published probability calculations. However, it is instructive to note a couple of examples of what is being published in the literature, as they, consistent with all published calculations (using different assumptions) known to the author, draw the same conclusion — which is the extreme unlikelihood that protein, DNA, RNA and cell construction could have initially occurred via natural phenomena.

The main example used here will be the probability of forming a single protein. Before getting into calculated results, a few relevant parameters and other considerations are listed:[78]

1. There are over 300 types of amino acids, most or all of which might have been present in primordial soups if conditions had allowed amino acids to form. Only 20 (plus two more for special cases) of these 300 types are used in life (i.e., in building proteins). So, to have life, the selection process for building proteins must have been highly discriminating. So, what discriminating natural phenomena process existed to select the right amino acids to use?

2. Amino acids generally come in two structural (i.e., isomeric) forms, left and right handed (i.e., mirror images of one another), usually in 50:50 ratios based on experiments using non-life components. Both forms have the same molecular formula. But only left-handed amino acids are used in biological proteins. So, what discriminating natural phenomena process existed to select the correct structural form of amino acids to use?

3. Amino acids must bond in the correct order using only specific peptide bonding — combining the carboxyl end of one amino acid to the amino end of the next amino acid, or the protein will not form or function properly. A primordial soup would have contained many other kinds of molecules, many of which would have bonded with any amino acids present, making them unavailable for making proteins.

4. Amino acids require an energy source for bonding, so a separate energy converter (mechanism) must have been in place as a means of converting sunlight (or other energy source) into usable energy. Where would the energy converter, which would have required energy itself to have been constructed, have come from? Is this is a Catch-22 situation? Also, energy conversion in cells requires protein catalysts — so where would the proteins have come from needed to construct the first energy conversion mechanism?

5. Any protein constructed without the protection of a cell membrane (which itself includes proteins in its structure) would disintegrate if exposed to an atmosphere containing oxygen, would disintegrate if exposed to UV rays from the sun (if there was no oxygen present to form the protective ozone layer) and might

disintegrate in water (due to hydrolysis or certain other chemicals present).

6. Natural selection (i.e., evolution) cannot be invoked at the pre-biotic level. The first living cell must be in place before natural selection can function.

Before citing specific statistical calculations, it is instructive to note what is considered by mathematicians as the statistical equivalent of zero. The threshold for eliminating chance as a cause of anything has been suggested by mathematician E. Borel as 1 in 10^{50}.[79] As will be shown, using this suggested threshold eliminates random chance from having been the cause of even a single protein being formed. And thousands of different proteins are needed for life. The author is unaware of anything in the literature that challenges Borel's threshold of 1 in 10^{50} as being the statistical equivalent of zero.

Proteins range in size from containing 50 to over 30,000 amino acids. Consider a small protein of 100 amino acids and refer to item 1 above regarding the amino acid sequence in a protein. To keep things simple, consider that only the 20 amino acids used in living cells are available for selection rather than all of the 300 different kinds of amino acids.

Since each position in the protein sequence can be occupied by any one of 20 amino acids, the probability of all of the 100 amino acids in the protein sequence being correct is $(1/20)^{100}$ which is equal to 1 in 10^{130}. This by itself is the statistical equivalent of zero (since the probability is even more remote than 1 in 10^{50}).

> **Note:** Of all the possible sequences of amino acids in a 100 amino acid molecule (10^{130} in the above example), usually only one forms the desired structurally correct and properly folded protein molecule. Most of the rest

are amino-acid chains that are either useless or potentially harmful to the cell.

Next consider the additional problem that all amino acids selected
must have a left-handed configuration (see item 2 above). Obtaining a
protein of 100 left handed amino acids (from a mixture containing both
mirror image isomers of amino acids) is analogous to flipping a coin 100
times and getting 100 heads (or 100 tails). This probability is 1 in 10^{30}, an
exceedingly small probability (approaching zero), which has never been
observed.[80]

So, the total probability of a 100 amino acid protein being formed,
considering only the above two factors (sequencing and the correct
structural isomer) is the product of the two probabilities, as follows.

Probability of random events resulting a single 100 amino acid protein needed for a particular purpose in a cell, considering only sequencing and isomer requirements, $=: (1/10)^{130} \times (1/10)^{30} = (1/10)^{160}$ (i.e., 1 in
10^{160}) which is a far lower (i.e., more remote) probability than what is
considered the mathematical equivalent of zero.

And the actual probability is even more remote than the above, as
other factors affecting probability (e.g., presence of amino acid types not
associated with life, amino acids reacting with other compounds present) have not been included. Clearly there is no need to include them
since the probability is already lower than the statistical equivalent of
zero. And most proteins are far larger than the 100 amino acid example
discussed above which would dramatically further reduce the odds. And
of course, the above considers only a single protein. A cell has many
thousands of different proteins that all must exist for a cell to function
properly.

Astronomer and Director of the Cambridge Institute of Astronomy,
Sir Fred Hoyle (originally an atheist) has stated: "Life cannot have had

a random beginning... The trouble is that there are about 2000 enzymes (which are proteins), and the chance of obtaining them all in a random trial is only one part in$10^{40,000}$, an outrageously small probability that could not be faced even if the whole universe consisted of organic soup."[81]

Well known astronomer Carl Sagan has cited a similar example of a 100 amino acid protein having a probability of occurring being 1 in 10^{130}, which agrees with the separately authored example developed above regarding sequencing. His conclusion was that, however peptides (i.e., proteins) were first constructed, it was not by random assembly.[82]

In a separate publication, one of the simplest peptides (i.e., protein) is discussed which is only 32 amino acids long. The probability of it forming randomly, in sequential trials, is approximately 1 in 10^{40}. That's "one" followed by 40 zeros.[83]

> **Note:** By sequential trials is meant that the molecule could have evolved stepwise, over time, vs. all components coming together in the correct order and structure at a single point in time (for which the odds are even more remote).

Prof. Stephen Meyer separately provides the following discussion which supports the above described analysis: To get tertiary (i.e., three dimensional) structure in a protein (required for its functionality), at least 75 amino acids or so are needed. To get a protein molecule to form by chance, first the right bonds between the amino acids are needed. Second, amino acids come in right- handed and left-handed versions (i.e., isomers) and only left-handed ones will work. Third, the amino acids must link-up in a specified sequence, like letters in a sentence.

Run the odds of these things falling into place on their own and the probabilities of forming a rather short functional protein at random would be one chance in 10^{125}, which is essentially zero. And that would be only one protein molecule. A minimum cell would need hundreds of different kinds of protein molecules.[84]

Since every living cell is far more complex and ordered than a single protein, it is essentially impossible that even the simplest form of life could ever have originated by chance. Even the simplest replicating protein molecule that could be imagined has been shown to have a probability (of occurring spontaneously) of essentially zero.

Biology professor F. Salisbury calculates the probability of a typical DNA chain developing from random events to be one in 10^{600}.[85]

Readers interested in additional probability perspectives may want to read the Back to Genesis article "The Mathematical Impossibility of Evolution" by Dr. Henry Morris (Founder of the Institute for Creation Research) which focuses on the probability of a multi-component system having evolved from a simpler system via evolution, given that most mutations are harmful to a system and result in degradation, disorder, and even a loss in functionality. Experience with purposely mutating cells shows that only about one mutation in a thousand actually results in some improvement to a cell.

While Dr. Morris did not describe a specific example, one can speculate that the acquisition and storage of energy in a living cell might be a great example. That is, the earliest cells were thought to depend on an influx of chemicals such as methane and sulfur as a source of energy — used directly by the cell. The development of photosynthesis (for plants) and the Krebs Cycle represent many subsequent iterations in developing the current cellular systems of energy acquisition and storage, with many beneficial mutations required to evolve such current systems.

So, the question is how likely is it for a sequence of many beneficial mutations to have occurred in the same system when the probably of even one having occurred is remote and given the high probability that many, if not most, of the far greater number of other mutations occurring in the same system were detrimental. The conclusion of Dr. Morris' quantitative analysis is that evolution by mutation and natural selection is mathematically and logically indefensible.[86]

Even if this wasn't the case, the initial question of life (from non-living matter: abiogenesis) still wouldn't be resolved as mutation and natural selection can only work on living organisms, which begs the whole question of the origin of life.

To help readers better understand the improbability of the information stored in DNA occurring by chance / natural phenomena, some authors have offered an analogy based on the 3 billion letters in DNA molecules (i.e., consisting of a mix of the four types of base pairs), noting their use in storing more than the amount of information embedded in 10 sets of Encyclopedia Britannica. They compare this to the likelihood of putting a mix of 3 billion letters (of 4 letter types, A, C, G, and T) in a bucket, withdrawing them three at a time (mimicking the codons in a DNA molecule), and expecting the resulting sequence to be an understandable properly punctuated decipherable set of 10 Encyclopedia Britannica (or something equivalent) — even if this process was repeated every second for a billion years.

The probability of this ever happening seems beyond human comprehension. Could someone even expect the much easier task to occur of such letter withdrawals sequencing into a novel such as "Gone with the Wind" (or something equivalent), which has less than 500,000 words (i.e., about 3.2 million letters) — which is only about 0.1% of the information that a human DNA molecule contains.

Keep in mind, as mentioned previously, that there is no chemical reaction or bonding driving force for a DNA molecule to naturally sequence itself with base pairs (i.e., letters) in any particular way. I.e., base pairs do not naturally interact or bond directly with one another. The base pairs are only held in place by their connection to DNA backbones (i.e., side rails) and not by any attractions or bonds with one another.

Famous ethologist, Richard Dawkins, has offered a counter-argument that statistical data based on totally random events may not be the correct way to think about the origin of life. It might be that random events eventually created a new better molecule, which then remained stable and was subject to random events that eventually created an even better (longer more complex) one. That would be akin to a random selection of letters in creating a book until the correct first letter was found (that being, e.g., the "S" in Scarlett O'Hara in the book *Gone With the Wind*), then that letter remaining fixed and random selection then occurring to find the second letter (that being the letter "c" in Scarlett), etc.

If natural phenomena worked the same way, that would improve the odds of natural phenomena causing the origin of life; however, the odds are still statistically zero. And that doesn't take into account that the first letter (with an analogy to a biomolecule) is unlikely to remain stable. Back in the prebiotic world, that first letter might be akin to random events somehow starting the creation of a nucleotide (i.e., building block of RNA or DNA). Such a molecule would not have been created anyway if the surrounding environment was oxidative.

Further, it is highly unlikely the nucleotide would have polymerized into an actual RNA or DNA molecule, as polymerization reactions don't normally occur in the presence of impurities (or at least don't occur with consistent quality or reproducibility). Even if such impurities were absent, the various forms of RNA do not form spontaneously in a mixture of nucleotides, but are only synthesized with the assistance of highly specific proteins. So, where would the proteins have come from?

And, finally, even if bits of RNA or DNA somehow formed, they would have been quickly destroyed by any UV radiation present. And there would have been plenty of UV radiation in the earth's atmosphere and surface as the earth's ozone layer would not yet have formed. Therefore, "the chance creation of proteins and DNA invokes extremely implausible chemistry."[87]

So, the number of factors to consider in RNA/DNA synthesis is large, — and all published analyses known to the author, for these and other reasons, conclude that the statistical probability of natural events causing the origin or life is essentially zero.

A confirming perspective is offered by the development of chemistry and physics based mathematical models of the Origin of Life. Several decades ago such models required assumptions about specific chemical reaction pathways and so did not have much impact on philosophical thinking (since evolutionary chemical pathways are not known for sure).

However, more recently, models have been developed which don't require assumptions about chemical reaction pathways; rather they only depend on specifying the initial chemical composition (e.g., chemicals in the primordial soup) when life supposedly began and then the complex composition that chemicals have in living cells. Using calculations, based in part on thermodynamics, and using high speed computers (to simulate the billion or more years of chemical evolution), the probability of chemical evolution having been the Origin of Life has been estimated.[88]

These results are in the same ballpark as some of the earlier examples and perspectives noted in this book, which is that the probability of chemical evolution leading to the origin of life is essentially zero. Ilya

Prigogine, the Nobel Prize-winning thermodynamacist, publishing on the probability that life occurred spontaneously, stated:

> The probability that, at ordinary temperatures, a macroscopic number of molecules is assembled to give rise to the highly ordered structures and to the coordinated functions characterizing living organisms is vanishingly small. The idea of spontaneous genesis of life in its present form is therefore highly improbable, even on the scale of the billions of years during which prebiotic evolution occurred.[89]

Since the accuracy of almost every calculated result can be challenged, it is not of value to cite more examples in this book. However, despite the different assumptions used in different published examples, all of the published probabilities known to the author are in a mathematical ballpark indicating the odds of the creation of life from randomly occurring activities are infinitesimal (i.e., the statistical equivalent of zero).

If the calculated results in the above examples (and in many others in the literature) are off by a factor of a billion or even a trillion, the conclusions are the same. The author has not yet seen any published article that presents a rational explanation, with reasonable probability of having occurred (i.e., odds better than 1 in 10^{50}), as to how the first living cell (or even one averaged sized protein molecule) could have been created by naturally occurring chemical and physical events.

Sir Fred Hoyle concludes this section of the book with a quote that: "Once we see, however, that the probability of life originating at random is so utterly minuscule as to make it absurd, it becomes sensible to think that the favorable properties of physics on which life depends are in every respect deliberate.... It is therefore almost inevitable that our own measure of intelligence must reflect... higher intelligences... even to the limit of God... such a theory is so obvious that one wonders why it is not widely accepted as being self-evident. The reasons are psychological rather than scientific."[90]

> **Note:** The use of psychological here means the refusal to admit to the possibility that a God might exist.

CHAPTER 10

Attributes of Higher-Level Organisms, Including Humans

Objective: As complex as a single living cell is, examine the additional complexity that must have happened via natural phenomena (if that theory is correct) in developing high level organisms containing trillions of cells — including attributes such as emotions, consciousness, and morality in humans.

As daunting as it is to try to understand (and accept) how a complex single living cell, including its DNA instruction set, might have formed through random events in nature, it is even more daunting to understand how the anatomy and physiology of living cells and tissues then evolved with incomprehensible further complexity to accommodate the additional functions and capabilities of high-level organisms, including humans.

So, contemplate the additional functions and features needed for different kinds of cells to work and interact together for the good of a 40 trillion-cell human organism, to allow it to grow, reproduce, and create new things, and to keep it, at least for awhile, from dying and disintegrating into dust. How could evolution have caused the development

121

of the huge number of complex subsystems needing to exist and work together in an organism when the huge majority of mutations to a cell (the source of evolutionary actions) are destructive to the cell?

So, while the main focus of this book is on the origin of the first living cells, some content is included here on the larger context of "life" — that being the additional aspects of life associated with humans and, in some cases, special capabilities of some animals.

Structure and Function

Questions that come to mind include:

1. **What evolutionary principle can explain how brain cells (i.e., neurons) developed so that they can, in real-time, generate new connections to other neurons and/or alter existing connections to other brain cells based on incoming sight, hearing, smell, touch, and taste electrical signals?**

How about neurons making new connections based on conscious thought (without sensory organs providing signals)? So, brain tissues can wire and rewire themselves. No known human developed hardware product or system is known that can wire or rewire itself. There are physical examples (i.e., computers) that were intelligently designed to automatically make adjustments to their internal logic pathways as new information comes in. However, such internal adjustments are done in software, not hardware.

> **Note:** Neurons and the axons, dendrites and synapses that connect them to other cells are a form of hardware.

Perhaps the most widely known example of a system with self-adjusting application logic pathways accomplished in software is IBM's supercomputer Watson, utilizing artificial intelligence (AI) and sophisticated analytical technologies to quickly search and analyze informaticn from multiple databases.

Watson eventually collected enough information and analyzed it fast enough to beat TV's Jeopardy champion, Ken Jennings, and one other former champion in a trivia match competition. But this system (Watson) required the combined creative intelligence of about 25 engineering ard computer software experts over 4 years to develop computer programs to allow the software to adjust its internal logic in certain limited ways.

Another example of self-adjusting software are computer based chess games which can be programmed to "learn" from experiences in playing chess matches against human chess players (or other computers) and become "smarter" after each match.

> **Note:** IBM's AI based "Deep Blue" computer was developed by experts to play chess and ended up beating world champion Garry Kasparov in 1997. So, other than life itself (which is still up for discussion), the only known physical systems utilized by mankind that are capable of partially adjusting, reconfiguring, or reprogramming themselves to achieve a directed objective (e.g., to become smarter), based on new information, have been built by intelligent creators (i.e., humans).

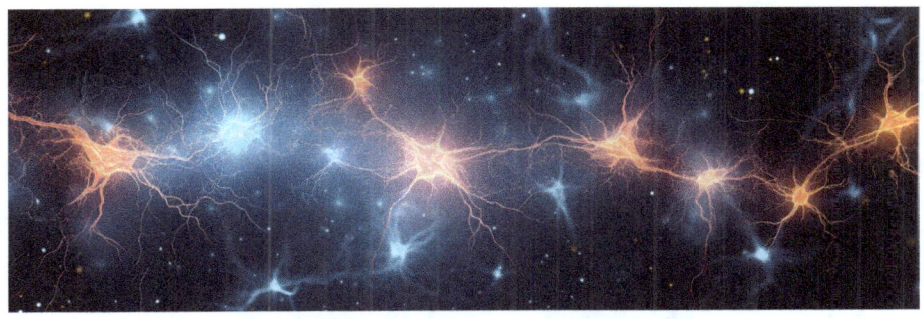

FIGURE 34: INTERCONNECTED NEURONS IN THE BRAIN

Further, does anyone think that as man-made computers become ever more powerful that they will eventually attain the equivalent of consciousness? Even if they do, it will have been accomplished by creative design from intelligent computer scientists (and not by random changes to computer software).

From a different perspective, humans have the maximum number of brain cells they will ever have (about 80 billion) between birth and about age 2. By adulthood, over 50,000 brain cells die every day.

> **Note:** While the number of brain cells (neurons) is decreasing over time, neural connections between cells increase for much of a person's active life.

Despite the loss of so many neurons every day, the brain keeps functioning continuously with no personal behavioral or memory changes noticed by others occurring on a day to day basis. Elegant complex neurological connectivity mechanisms must be involved. Compare this to the gigabytes of memory in computers when the loss of even one memory cell, depending on what software utilizes it, can potentially cause the computer to crash.

2. **What evolutionary principle can explain how bones developed to meet the many needs of an animal's structural support (plus their use for other purposes)?**

The newborn baby has 350 bones, many of which gradually fuse together into the 206 bones carried by most adult humans. Many of a baby's bones are soft and pliable, needed, e.g., to facilitate a baby passing down the narrow birth canal during natural birth. Bones are connected so as to allow maximum freedom of body movement. Consider hands and feet, each of which contain about 26 bones. How could evolution (i.e., random chance) have developed such an intricate sophisticated foot and hand design, unmatched by anything mechanical that humans have ever developed? Think of the creative development that has occurred with commercial robotic arms and pincers, such as claws, jaws, and grabbers, — which are still clumsy and limited compared to human arms, elbows, wrists, and hands.

After birth, bones grow continuously. They replace 100% of their cells every year during a body's early years, slowing to about 20% per year later in life. Bones eventually harden into strong lightweight material (1/3 the density of steel) that can lubricate themselves, require no shutdown time, and repair themselves when damage occurs.

Many bones take on auxiliary seemingly unrelated functions such as serving as the home for independent "marrow" tissues that, collectively, manufacture up to 500 billion new blood cells per day. How did evolution select the cavities inside some bones for this unrelated purpose (if evolution was the cause)? That is, marrow inside certain bone cavities manufacture three kinds of cells (red blood cells, white blood cells, and platelets) that are unlike bone cells.

Another unique mission of certain bones regards the three bones (known as ossicles) in each ear that serve to transmit vibrations from the eardrum to the nerves in the cochlea so that an animal can "hear." How could evolution have developed this complex system — which involves converting and transporting different kinds of energy (sound waves to bone vibrations to fluid motions to electrical signals) — using different kinds of tissues, all synergistically working together at the same time to allow hearing to occur?

Also, where did the 3 bones come from? They are independent of and not connected to any other bony structures in the skeleton. And it's even more complicated than this regarding sound quality, as a long Eustachian tube extending from the middle ear (where the ossicles are located) to a different part of the head exists (to equalize pressure between the middle ear and atmosphere) in order to optimize transmission of sound waves within the ear.

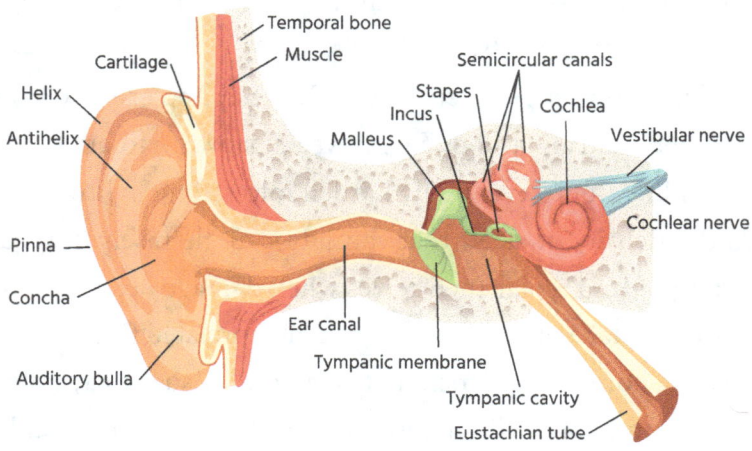

FIGURE 35: ANATOMY OF HUMAN HEARING
(THE MALLEUS, INCUS, AND STAPES ARE VIBRATING BONES)

This book has already noted the many thousands of chemical reactions that occur inside cells and organisms utilizing chemical energy. When adding the sequence of multiple transformations of energy forms required for hearing and the need for multiple integrated subsystems to simultaneously work together, it seems undeniable that a purpose, plan, and creative acts must be responsible.

Remember: Evolution has "no end in mind." Each of the trillions of DNA molecules in the body has the instructions for constructing the complex anatomy associated with hearing. How could natural phenomena have caused this?

Regarding joints, human natural cartilage on a bone's surface continues to be the best-known bearing surface for joints as there is almost no friction involved (due in large part to self-lubrication). All synthetic joint materials developed by creative mankind (for use in joint replacements) have higher friction rates than cartilage — resulting in higher wear rates due to tiny particles of synthetic material being released from the joint surface which can eventually cause inflammation of the joint.

> **Note:** The life span for a natural biological joint for most people (unless joint injury or disease sets in) is the same as the life span of the organism (of which the joint is a part) which is over 70 years; the published expected life of a synthetic joint replacement is on the order of 15-20 years.

So, intelligent mankind has yet to develop anything as effective for joints as what the body naturally provides.

A question is, what structural material made by creative man has as many beneficial qualities for the body with minimal side effects as bone? That is, no scientist has yet discovered or developed a material as well-suited for the body's support needs as bone, occupying 20% or less of the body's weight.

To this day, a preferred material in replacing small or modest amounts of damaged or dead bone is "bone grafts" (often using bone taken from an individual's own upper pelvis). Could it be that chance random acts of evolution produced materials (whether bone or other tissues) that creative man has not been able to improve upon? Or was purposeful creative design involved?

> **Note:** Chemists and engineers have access to the same elements and materials (plus a lot more) as do biological systems.

3. **What evolutionary principle can explain how cells, in addition to managing their own metabolic and replication functions, became able to interact and control the operations of other dissimilar cells and tissues within an organism?**

For example, cells in the pituitary gland manufacture hormones that control growth throughout the rest of the body; other kinds of cells in organs produce products that help control the body's blood pressure and blood sugar level.

Regarding blood sugar (i.e., glucose) control, beta cells in the pancreas increase production of the protein hormone insulin if the blood glucose concentration gets too high, such as after meals. Insulin protein molecules then attach to specific receptor sites on the cell membranes (analogous to a lock and key relationship) of the trillions of cells in the body — which then prompts the cells to allow glucose in the blood to enter the cells (which, in turn, lowers blood sugar level). Insulin also prompts liver (and muscle) cells to store (as glycogen) any excess glucose that exists in the blood.

> **Note:** Glycogen is an insoluble glucose polymer representing the storage form of glucose.

If blood glucose gets too low, alpha cells in the pancreas produce more of the protein hormone glucagon which then prompts special cells in the liver (and muscle) to release glycogen, thus raising blood sugar level. So, somehow, the ability to:

a) internally continuously monitor blood glucose concentration,

b) produce two different hormones by two different types of cells in response to continuously monitored blood glucose concentration,

c) employ the use of two different organs of the body,

d) utilize highly specific receptors on most of the body's individual cell membranes (responsive only to insulin), and

e) utilize specialized receptor sites on liver (and muscle) cells (responsive only to glucagon) — all need to be in place and working together to accomplish blood sugar control.

Glucose control is actually even more complex than described above, utilizing additional hormones, such as somatostatin, which slows the production rate of insulin and glucagon when necessary. How could such a critical complex system have come into being via evolution (which has no end in mind as changes occur via mutations)? This is a far more sophisticated and complex control system than a car's cruise control which only came about from the creative intelligence of automotive engineers.

So, organisms are known to "adapt" to their external environment over time (e.g., via natural selection and survival of the fittest) but how did cells acquire the ability to help "control" certain activities of different remotely located cells in their external environment?

4. **What evolutionary principle can explain how living things can diagnose and repair (i.e., heal) themselves from a wide range of injuries and other maladies?**

The process of healing/repairing is a sequence of multiple phases involving multiple body parts and functions that is multifaceted, complex, interactive, and highly coordinated. Often the self-manufacture of new body tissues, to order, is needed.

As an example, one of the many body functions involved in healing is the immune system, especially when foreign microorganisms are present. White blood cells, in particular, located in the blood, lymph nodes, and elsewhere, are a key component of the immune system. Specialized white blood cells called lymphocytes can distinguish between the over 200 normal different kinds of cells in the body and the foreign entities that they should attack. B lymphocytes (i.e., B cells) produce proteins called antibodies that attach to foreign invaders (and/or the toxins they generate) so other immune system cells, such as T cells, can destroy them. Could such a complex system have evolved from random natural events in nature?

In contrast to an organism's ability (in most cases) to heal itself, humans, as intelligent as they are, are not known to have created a complex physical system capable of automatically repairing itself. Has an automobile ever been developed that can fix its own flat tires? Has a computer system ever been developed that can, by itself, identify and fix the "bugs" that exist in its programmed operating system code (e.g., the logic errors that can cause a computer to crash)? If someday they can, it will have been the result of creative human intelligence.

When humans develop a complex system, such as an automobile or computer, a large infrastructure network of intelligent people resources (many highly educated) are usually organized and trained to help support the system — including diagnosing and fixing problems, some of which occur only rarely. New hardware parts needed for repair often have to be ordered and obtained externally. How would natural phenomena and evolution have developed the equivalent of such support systems in creating self-contained automated problem diagnosing and solving capability within living cells, in which all the parts needed are already available or can be manufactured by the cells themselves or by other cells within the organism?

5. **What evolutionary principle can explain how humans can be the same in so many respects (e.g., every undamaged cell containing**

DNA in an individual human has identical DNA and uses the standard Krebs cycle to capture and store energy) but, at the same time contain uniqueness among each of the billions of humans on earth (possibly excepting identical twins)?

These unique traits include things such as as a human's fingerprints and the eye's iris structure. Remember, not even the iris structures in the two eyes of the same human are the same.

> **Note:** The above unique traits are not thought to be related to DNA, but to other unknown organism development activities.

6. What evolutionary principle can explain how strings of biochemicals, even if created by natural random events and evolution, further evolved via random events such that the organisms they were part of developed languages and speech as a means of communicating with one another?

7. What evolutionary principle can explain how the different species of animals ended up with DNA (after presumed billions of random chemical reactions and mutations occurring over time) that represented the blueprint (i.e., recipe) for constructing two different versions of their species, one male and one female?

The two versions contain different sex organs (which presumably randomly evolved) that are required to connect to and interact with one another to enable reproduction. I.e., males require females for reproduction, and vice versa (so each version requires the other). Note that DNA evolves via mutations, which have no "end in mind," with most being detrimental. So, how did DNA evolve via random events so as to direct the construction of two different versions of complex organisms, with both versions becoming available at the same time and each requiring the other in order to sustain the species?

This is not a problem for creationists (with God creating Adam and Eve); not so for evolutionists.

Similar challenges exist in understanding how evolution might have resulted in other internal complex biological entities and systems of most animals, including the outer layer (e.g., skin), various internal systems (e.g., muscular, endocrine, reproductive, digestive, immune), and the five senses. The question is how did the cook book (i.e., DNA) come about that orchestrates the construction of all these entities and subsystems and how they communicate and function together for the benefit of the entire organism.

> **Note:** Recall that the DNA in each cell contains all the instructions for the whole organism, regardless of what cell it is in. So, how could such a cook book containing 3 billion letters of purposeful intelligible decipherable code have developed via random acts of nature?

The human body is like a highly organized symphony of millions of parts and subsystems working together to perform highly coordinated extraordinary beneficial outcomes. It took brilliant intelligent creative composers (e.g., Mozart, Beethoven, Brahms, Bach, Chopin, Strauss, Gershwin, Williams) to write the symphonic music (i.e., instructions) for a traditional 80- to100-member symphony orchestra. And it takes a conductor to choreograph, arrange, and lead the playing of the symphony.

So, how could random chance, the lack of brilliant composers (i.e., scientists), and lack of a conductor (i.e., divine agent) have produced a system of far greater complexity (than a symphony), requiring far more integrated working parts, subsystems, and coordination?

Emotions, Consciousness, Spirituality, Morality, perhaps a 5th Dimension

Questions that come to mind here include:

1. **What evolutionary principle can explain how a network of cells and tissues developed the ability that allows mankind to feel and display various emotions and feelings, such as empathy, sympathy, humility, anger, anxiety, compassion, contentment, jealousy, envy, appreciation, competitiveness, remorse, mercy, frustration, love, physical attraction, patience, kindness, peace, faithfulness, and joy? Also, why are some of these emotions unique (as best as is known) to only one organism (i.e., humans)?**

    Note: Some emotions are reported as observed in certain animals, such as fear, anxiety, and joy in family pet dogs; empathy in primates.

2. **What evolutionary principle can explain how humans developed the means to differentiate "right" from "wrong" and developed discernment in how to behave — which sometimes differs from natural biological instincts?**

For example, if one is hungry, one's biological instinct might drive one to steal a loaf of bread (much like Jean Valjean did in the world-renowned Victor Hugo novel, *Les Miserables*), which acceptable social and moral behavior (and one of the Bible's Ten Commandments) would tell us not to do.

FIGURE 36: MORAL STRUGGLE BETWEEN GOOD AND EVIL

Rev. Glenn McDonald, in his blog to his subscribers of Jan. 7, 2022, expands on this topic, summarized as follows:

> Most atheists and agnostics are perceived as valuing many of the attributes most people have, such as the need for social justice and educating the poor. However, when asked, most atheists and agnostics (incl. secular scientists) are hard pressed to justify such values according to their belief system. Judeo-Christian believers can point to their religious faith for such justification. Atheists and agnostics cannot.
>
> So, what causes the desire for people "to be and do good." The West's secular-minded universities, institutions, and philosophers have often wrestled with this question.
>
> Since secular materialism denies any ultimate basis for right and wrong, good and evil, truth and falsehood, — then non-religious human cultures have to

come up with an explanation as to why they feel the need to make moral choices.

There would seem to be two possibilities, evolutionary biology and social custom. However, neither biology nor social custom can provide firm reasons why it's wrong — not just in certain situations, *but always wrong* — for the rich to cheat the poor, for adults to sexually abuse children, and for people in power to take whatever they want for themselves. Atheists and agnostics generally believe these things are wrong but such convictions don't come from secular materialism. A philosophy that declares there is no God, no natural law, and no final accountability to a Cosmic Judge could not have come up with values like social justice and universal human rights.

But Judaism and Christianity certainly could — and they did.

If religion doesn't provide final answers about morality and meaning, where do such values come from? Goodness, kindness, and human rights didn't arise by chance, by some roll of the Darwinian dice.

So, most people embrace honesty, integrity, and treating others with dignity. But an agnostic or atheist-based belief system can't explain why such values are inherently superior to lying, cheating, and ripping other people off. Judeo-Christian believers can explain these values as they draw upon the Judeo-Christian conviction that a personal God, who is both good and just, is ruling the cosmos.

Human rights make sense because the Bible's book of Genesis indicates that all human beings are made in

God's image. Working for justice makes sense because God himself cares about justice and will make sure that "justice for all" is finally accomplished in this world or the next. So, most human's love of goodness makes perfect sense *only if* there is a good God who put such love into their hearts.

4. What evolutionary principle can explain how most humans believe that there is design in the makeup of their bodies and that there is "purpose" to their lives?

The Bible teaches that each life has value and each life is significant in terms of the purpose of God. That is, in the course of time and relationships, people have a purpose to fulfill in terms of the grand design God has planned. The Bible's Psalm 57: 2 states: ".... to God who fulfills his purpose for me." Some individual humans may not know their "purpose" in life but God, never-the-less, according to the Bible, has a purpose for everyone.

If only natural phenomena events were involved in creating life, humans would undoubtedly not exist in their present form if life's evolution processes were to be repeated, as evolution does not have any particular goal or "end-in-mind." So, evolution by itself, would not give a human a personal sense of why he/she exists on the earth and, if starting over with the first life form (presumably single cells), would undoubtedly not result in the present anatomical form of humans.

The Bible says that God made man in his own image (Genesis 1: 27) and, at one point in time, his son Jesus came to earth in human form (who looked physically just like a human). So an "end in mind" must have existed when creating life and only creative intelligence, and not evolution, could have caused this.

5. **What evolutionary principle can explain how salmon can leave their place of birth (in fresh water streams), travel thousands of miles away, live in salt water, and then return, years later, without Global Positioning Satellites (GPS), to their original birthplace to provide for the next generation of life?**

FIGURE 37: SPAWNING SALMON, RETURNING TO THEIR BIRTHPLACE

Some scientists have theorized that salmon memorize the geo-magnetic signature of their birth place before leaving their place of birth and then, years later, home in on that memorized geo-magnetic signature.

Not only are such theories unproven, but they are also problematic, since the forces that determine the earth's magnetic field are constantly changing. For example, the magnetic north pole is believed to be constantly moving east (from northern Canada to Siberia) at about 30 miles per year due to changes in the molten iron distribution in the earth's outer core.

So, the geo-magnetic fingerprint of a specific location at one point in time will be different several years later. Of course, the above phenomena is not unique to salmon. The migratory behavior of many birds and other animals appears to be similar.

If a human, living thousands of miles from their birthplace, were asked to return to their birthplace, the journey would be unlikely to successfully happen without the use of creative tools that man has developed, such as compasses, maps, and GPS. Isn't it reasonable to believe that some animals, tasked with the same challenge, were endowed by a creator with the means to accomplish their migratory journey without such tools? Such capability must have been the result of a purpose, plan, and creative design.

6. **What evolutionary principle can explain how man can desire and invite a perceived holy presence into his heart and mind (i.e., the holy spirit — which is believed by Christians and certain other religions to be one of the forms of God) and allow that spirit to guide his life's behaviors — which results in a "peace of mind" in life that surpasses all understanding (ref. : Philippians 4: 7)?**

Further, what evolutionary principle can explain how man acquired faith; i.e., trust in the assurance of things hoped for and the conviction of things not seen (ref. : Hebrews 11: 1)? For most members of religions, this has manifested itself, in part, in prayer — which is a solemn request for help or expression of thanks- addressed to God or an object of worship (e.g., deified ancestor). Note that a National Day of Prayer has existed for 70 years in America (observed in May), and that many decades of American Presidents have called for prayer for our nation.

7. **What evolutionary principle, mutation, or other natural phenomena can explain how Jesus, the accepted son of God by a large segment of humanity (and stated as such in the Bible), came into being without the benefit of a sperm cell (with a sperm cell needed to provide half of the DNA required for a fertilized egg and also required in order for a mammal to produce a male offspring)?**

As stated in the Bible, Jesus was born of a virgin. The birth of Jesus represents a unique, novel, and intelligent intentional act of a Creator as it did not begin via naturally occurring events.

The list goes on. The point is that it is one thing, as complex as it is, for cells and tissues to come together to perform basic metabolic and reproductive functions inside complex semi-permeable cell membranes and walls. It is yet another to understand how the anatomical and physiologic make-up of an organism can accommodate the knowledge building, discernment, emotional, and spiritual functions that humans possess.

Given the existence of mankind (who has never been able to create life from non-living matter in the laboratory), and knowing that all known complex physical manufactured systems in the world have required mankind's creative intelligence, it has been logical and sensible for hundreds of millions of people to accept and understand that an even greater creator must exist that has provided for all the above listed functions, and at least guided the events that led to humans and their many wondrous capabilities.

Other perspectives have been published in support of (or challenging) the creation theory. Love will be used as an example, with perspective offered by famous evangelist, Dr. Billy Graham from one of his daily published newspaper columns. He explains:

> God made us because of His love. On a human level, we know that love needs an outlet — i.e., love yearns to be expressed and shared. In a far greater way, God's love is the deepest expression of His desire to have fellowship with mankind. That is why he created Adam and Eve. And He created them in His image so that they would have the ability to love also — to love each other and to love Him. Our loving God had a compelling urge to cre-

ate humanity. His love was expressed in the creation of the human race.

So, God had purpose in creating the ability of humans to love (along with many other human characteristics).

> **Note:** When referring to God's creating man in His image, the "image" is a spiritual, moral, compassionate, and intellectual one — with man having consciousness, a soul, and capability of a relationship with Him-rather than God having a particular physical form (even though Michelangelo depicted God as a physical entity in human form in many of the scenes he painted on the ceiling of the Sistine Chapel in Italy's Vatican City).

A 5th dimension, 6th sense, etc. Of relevance also is that the laws of physics and mathematics do not preclude the existence of dimensions in addition to the four dimensions that most people know about, understand, and relate to (i.e., length, width, height, and time) — often referred to as "space-time." In fact, a team of scientists, led by Andrew Strominger in 1996, defined ten space-time dimensions that must have existed as part of the "Big Bang" and which must have existed since then to explain how gravity and quantum mechanics have coexisted.[91]

So, the laws of physics and mathematics allow for the existence of additional dimensions (i.e., more than four), even though some dimensions are perhaps not of interest to most people and may not have labels that most people are familiar with. Physicists already suggest such additional dimensions to explain the behavior of certain subatomic particles, including quarks, mesons, photons, neutrinos and gravitons (some of which exist in theory but have not yet been directly seen or measured). So it seems scientists sometimes practice faith also.

And, has anyone seen gravity lately (not the effect but the actual force)? Perhaps things like thoughts and emotions, which one can neither see, hear, touch, smell, or taste, can be thought of as additional dimensions. Perhaps there is a spiritual dimension. For example, most humans can identify with a "soul" that is part of their being. Most members of several major religions also believe in the "Holy Spirit" (claimed as one of three manifestations of God by Christians) that can be invited into one's being (symbolically the heart) and help influence human thoughts and emotions and help drive decisions on life behavior.

These are things that cannot be described in terms of the standard 4 dimension physical "space-time" world but which perhaps exist as part of a 5th or 6th (e.g., spiritual) dimension (or maybe it's an 11th or 12th dimension — given the physicist's current 10 space-time dimension model) where God's Holy Spirit operates. Humans may not be able to see God directly while alive on earth, but many, if not most, people can relate to a soul and spirit that can be "felt" but not directly seen in 4 (or scientist's 10) dimensional space-time.

And, while some consider the topic controversial, there is significant anecdotal evidence of a dimension known as extra sensory perception (ESP), i.e., a 6th sense, that some people claim to experience. This relates to information gained not via the recognized physical senses, but with the mind. It captures such capabilities as intuition and telepathy.

And what about the "conscious mind"? Mind and matter are two quite different things. Only humans possess "mind" — the capacity for self -knowledge and abstract thought.[92] So what evolutionary principle created the conscious mind?

Many scientists and philosophers are now concluding that the laws of physics and chemistry cannot explain the experience of consciousness in human beings. They sometimes refer to this non-material reality as the "soul," "mind," or "self." Humans have the capacity for self-reflection, representational art, language and creativity. Science can't account for this kind of consciousness merely from the interaction of physical matter in the brain. So where did it come from?[93]

Wilder Penfield (renowned father of modern neurosurgery), agrees with the Bible's assertion that human beings are both body and spirit. "To expect the highest brain (physical) mechanism or any set of reflexes, however complicated, to carry out all the functions of the mind, is quite absurd."[94]

Jesus described (in the Bible) the body and soul as being separate entities when he said "Do not be afraid of those who kill the body but cannot kill the soul" (ref., Matthew 10: 28).

Regarding Consciousness and Natural phenomena, the well-known scientist, theologian and philosopher, J. P. Moreland, comments: "How do you get something totally different — conscious, living, thinking, feeling, believing creatures — from materials that don't have that. That's getting something from nothing."[95] Darwinists / Evolutionists / Naturalists can't explain the origin of consciousness because they can't explain how you can get something from nothing.

Sound familiar? The agnostics and atheists (including many scientists) that believe in natural phenomena as the origin of life and humanity have a similar problem in accepting the Big Bang theory regarding the beginning of the universe (i.e., they don't accept theories requiring an intelligent creator or events when something was created from apparently nothing).

Note: Some scientists don't believe the Big Bang represents getting something from nothing. It is presumed the Big-Bang was preceded by the existence of a massive amount of energy and that the Big Bang was a transformation of energy into matter — perhaps in accordance with Einstein's famous discovery that $E = mc^2$ where E = energy, m = mass, and c = the speed of light. This equation was used in the development of nuclear power and nuclear weapons, showing that mass could be converted to energy. The Big Bang perhaps suggests the reverse reaction: i.e., energy being converted to mass, perhaps using the same equation.

Human beings are not only endowed with mind, but also with morality. Where did that come from? Morality is a form of law and moral law necessitates a law giver. The "ten commandments" are a form of moral law that people of several religions know and understand. Members of the Christian and Jewish faiths believe that those commandments were given to Moses and his people by God. The Bible says so.

As stated by E. Andrews in the book, *Who Made God?*: Why should our physical senses be able to detect God if, as Jesus declared (in John 4: 24), "God is Spirit, and those who worship Him must worship in spirit and truth"? Our five physical senses, no matter how much enhanced by scientific instrumentation, remain physical — developed to explore material and physical objects and processes, not spiritual ones.

The apostle Paul declares (Galatians 5: 22-23): "the fruit of the Spirit is love, joy, peace, long-suffering, kindness, goodness, faithfulness, gentleness, and self-control." These are abstract concepts and experiences that are not accessible to our scientific instruments. But they are undeniably real.

CHAPTER 11

Summary and Conclusions

> **Objective:** Summarize key points from earlier sections of the book. Summarize the well-known Watchmaker Argument and consider the possibility that the Origin of Life could be considered as a miracle.

The Watchmaker Argument

One argument for "creation" has been captured in a classic historical published example, developed by eighteenth century British theologian-naturalist William Paley, which is called the "Watchmaker Argument." It goes like this.

Suppose in walking along a path, one comes upon a boulder. If one is asked how it came to be there — one might be tempted to say that it could have been there forever, having been created by the natural forces of nature. But what if, instead, one comes upon a watch on the path. One would never think that it had always been there.

The watch must have had a maker and there must have existed a place and time in which the watch was designed and constructed. I.e., no one would ever conclude that a watch was the product of bits of dust, dirt, and rock being shuffled together via natural processes. Even if the natural processes were allowed to operate for a very long time, there would still be no rational hope for a watch to be assembled as a result of wind, rain, fire, UV radiation, lightning, gravity, atmospheric gases, and/or other natural phenomena.

FIGURE 38: ILLUSTRATING THE WATCH MAKER ARGUMENT

Yet, the complexity and capability of living organisms far transcends anything that exists in a watch. If a watch's complexity and capability demand an intelligent and creative maker, surely living organisms demand a Maker of far greater intelligence and creative ability.[96]

Dr. Hugh Ross, in arguing against any benefit of long time periods available in the origin of life, takes the watch example a step further. He begins by supposing that all the precisely formed parts of a watch (created from natural phenomena events- which is a near impossibility in itself) happen to be available at one time and are placed in a shoe box (simulating a cell wall) and exposed to mild shaking (simulating, e.g., environmental events such as earthquakes, blizzards, etc.).[97]

Even if the extremely improbable situation occurred that all the parts came together in the proper orientation at a point in time (and had not rusted or eroded over time from multiple collisions with other components), the shaking of the box would have to be instantaneously stopped so as not to have continued shaking cause the parts to disassemble. I.e., at any point in time, the mechanical and motion forces involved during shaking would result in a far higher probability of disassembly to occur than of assembly. That is, time favors disassembly (i.e., more random-ness) and not assembly. So, the precise timing of exactly when to stop shaking the box would itself require an intelligent "box shaker."

Summary of Key Conclusions

The simplest living cells in the world are still amazingly complex entities, with thousands of different compounds and chemical reactions involved in just maintaining the cells, with additional millions of other molecules involved in the ability of a cell to replicate. One wonders how did such a diverse mix of so many compounds ever find and assemble themselves inside a semi-permeable membrane controlling many com-plex processes (metabolism, growth, replication) and including control of what materials go into and out of the cells.

Clearly the number of components and associated complexity of operations inside even a single independent cell exceeds that of the most complex machines and systems ever built by the creative powers of man, such as commercial airplanes, high-rise buildings, and super comput-ers. Said another way, since the creation of civilization's most advanced machines and systems required an enormous amount of directed intel-ligence, what logic can explain how something even more complex (i.e., a living cell) came into being all by itself without any creative directed intelligence. It doesn't make any sense.

In looking at the structural (sequential and folding) requirements of specific proteins, or DNA itself, the odds of such entities being created naturally from random events and natural selection are essentially zero, even if such events occurred over a long time in sequential steps. Yet those scientists believing in the natural phenomena theory believe that time, chance, and the inherent properties of matter caused the origin of life, even though inherent properties of matter would not naturally cause molecules such as DNA to synthesize and certainly would not cause such molecules to become a warehouse of information needed for cellular activities.

The inherent properties of matter also cannot explain how non-material aspects of human life came into being, such as the soul, consciousness, and emotions. It is far more logical to believe that life, including both original cells and subsequent organisms, began by purpose, plan, and special acts of creation, just as has occurred for all complex systems devised by humans.

Isaac Asimov, prolific author, ardent anti-creationist, and a Professor of Biochemistry, has stated: "In man is a three-pound brain which, as far as we know, is the most complex and orderly arrangement of matter in the universe." It is much more complex than the most sophisticated computer ever built. Wouldn't it be logical to assume that if man's highly intelligent brain designed the computer, then the human brain was also the product of design?[98]

Biology professor Dean Kenyon has concluded that "nothing short of an intelligence could have created this intricate cellular apparatus. This new realm of molecular genetics is where we see the most compelling evidence of design on the Earth."[99]

The origin of life is a set of paradoxes. In order for life to have begun, there must have been a genetic molecule—something like DNA or RNA—capable of passing along blueprints/instructions for making proteins, the workhorse molecules of life. But modern cells can't copy DNA and RNA without the help of proteins themselves.[100]

So, which came first, DNA or proteins? They each require the other. Another paradox is that of manufacturing proteins. Proteins require enzymes to enable the protein's formation, but enzymes themselves are proteins. There are several other "chicken and egg" paradoxes regarding the origin of life (including the biosynthesis and use of lipids) — with additional examples beyond the scope of this book.

The origin of life requires all three of the following to exist: building blocks, external energy sources, and instructions. Building blocks (e.g., amino acids, the building blocks for proteins) would only have existed in primordial soups if the environment was reducing in nature — with recent evidence suggesting the environment was not reducing but neutral or perhaps even oxidative (as it is today). Recall that amino acids are not synthesized in an oxidative environment.

The presence of some oxygen (suggesting an oxidative environment) has been theorized as existing in early Earth's atmosphere, as it is the natural product of UV radiation acting on water vapor, both of which existed in the early atmosphere. Also, how otherwise could the ozone layer have formed, which uses oxygen as the starting raw material?

Oxygen has been shown to exist in the atmosphere of some celestial bodies where life is not known to exist. Also, external energy sources existed in earth's primordial times (e.g., UV light, lightning), but such sources are more likely to have been destructive in nature (particularly regarding macromolecules such as proteins and RNA) than supporting synthesis.

But perhaps the biggest dilemma for "naturalists" is an explanation of the origin of instructions (e.g., in DNA) for purposes including:

1) making complex proteins,

2) manufacturing macromolecules, and

3) assembling a super complex self-replicating organism.

In fact, since hard experimental proof is in short supply for either of the two prevailing theories of the origin of life, mankind might draw on other perspectives to compare the two theories. For example, experience (as well as the formal laws of information) indicates that large amounts of "information" invariably originates from an intelligent source, i.e., from a conscious mind or a personal agent. That is, intelligent activity is the ONLY known cause of the information needed in making a complex system starting from a non-living source.

Indeed, mankind's collective experiences affirm that relevant information, whether inscribed in hieroglyphs, written in a book, encoded in a computer program or what would be expected in any "information" that might be received from NASA's SETA search for extraterrestrial intelligence — always arises from an intelligent source.[101]

Noted British astrophysicist, Steven Hawking, is quoted as saying, "What is it that breathes fire into the equations (e.g., gravity) and makes a universe for them to describe? " That is, scientists seem preoccupied with the development of theories that describe "what" the universe is — and ignoring the question as to "why."

If maintenance and replication functions of the first single cell microorganisms are not complex enough, it seems unfathomable to understand how evolutionary processes could have enabled different types of cells to organize themselves into complex multi-trillion cell organisms — with such organisms capable of defending themselves against invading foreign microbes, repairing themselves, and, in the case of humans, developing a conscious state, moral code, and displaying such functions as joy, remorse, forgiveness, love, jealousy, envy and a host of other emotions.

Evolution theory is the interplay of two simple processes — random genetic mutations (i.e., accidental change in the genes) and natural selection (i.e., the survival of the fittest). Random mutations can have no "directional goal, such as the drive to greater complexity or functionality in living things. So, that leaves natural selection to impose directionality. However, natural selection can only select what is already present — it has no creative power.[102]

If the reader believes in the possibility of intelligent creative forces existing in a 5th (or greater) dimension (e.g., spiritual) that has influenced the origin of life and contributes to an individual's internal conscience that helps influence daily thoughts and emotions, they are in great company. Hundreds of millions of people, including many scientists and engineers, already do. Are they all wrong?

If all the complex physical entities built over time have been accomplished via the creative powers of man, why is it unreasonable to believe an even more creative entity (i.e., God), overcoming the natural tendency of matter and systems toward disorder, directed the creation of life (which mankind has never been able to mimic)?

A Leap of Faith

The truth is that both views (creationism and evolution/natural phenomena) are grounded in a leap of faith and both claim to be reasonable (although this book joins a multitude of others in questioning the reasonableness of natural phenomena as causing the beginning of life). The creationist places their faith in intelligent design, (paralleling in several respects what humans have intelligently created) and finds in this faith a reasonable explanation of life and its origin.

However, the evolutionist/naturalist also operates by faith: faith in the inexplicable and wholly random origin of something out of seemingly nothing (e.g., the universe via the Big Bang, consciousness, human emotions). Make no mistake, this is a leap of faith, an astounding assumption not based on known laws of nature.

There is also some level of faith in science believing that the descriptions of certain phenomena, seemingly in direct conflict, are simultaneously true, such as the wave and particle descriptions of light and electrons. For most members of many religions (e.g., Christianity, Judaism, Islam) faith in intelligent design as the explanation for the origin of life is far more reasonable than in faith that millions of improbable incidents of randomness and blind chance occurred, causing a live cell to develop that is far more complex than anything man has ever created.

> **Note:** To be clear, there are major philosophical differences between the different religions of the world (and wars have sometimes been fought over the differences). The only point here is that at least three of the major religions of the world (in terms of membership), including the top two (Christianity with over 2.3 billion members and Islam with over 1.9 billion as of 2021)[103] believe in a God that was responsible for the "origin of life." None of the major religions are known for advocating natural phenomena as the origin of life, although some religions are mute on the topic. E.g., while there is a definitive God directed Genesis in Jewish, Christian and Islamic teachings, there isn't one in Hinduism or Buddhism.

Said another way, the origin of life debate is ultimately about faith vs. faith. The atheist has great faith in extremely complex systems developing by chance, and the creationist has faith in a God who has designed the universe that humans live in and then brought forth life.

One wonders if naturalists/evolutionists/atheists/agnostics are more likely to believe:

1) someone can buy a single lottery ticket to each of ten different national power ball drawings and then win all ten jackpots

OR

2) natural phenomena resulted in the beginning of life

The odds are much better (i.e., more likely) for option 1.

Note: The odds of winning just one Powerball jackpot is about 1 in 300 million.

Recall also some of the previous examples cited where science had "faith" for long periods of time that what they were teaching was correct, despite lack of any direct proof, such as the earth being the center of the universe, the existence of black holes in the universe, and the existence of Higgs Bosons (known as the God particle). So, faith has sometimes been an important part of what scientists have concluded and taught. Sometimes, such faith has turned out to be based on flawed information and/or assumptions..

As noted by English physicist, Dr. Paul Davies, there is another aspect of faith to consider regarding science. "To be a scientist, you have to "have faith" that the universe is governed by dependable, immutable, absolute, universal, mathematical laws..." even though such laws are of an unspecified origin and why they work as they do is unknown.[104]

Take Sir Isaac Newton's Law of Gravity, for example, which is one of the great science discoveries of all time. Gravity is one of four known fundamental forces of nature. Because of gravity, every particle in the entire universe actually attracts every other particle. But what really is gravity? How is this force able to cause objects to physically interact with each other over vast distances of seemingly empty space? And why does gravity exist in the first place?

Science has not been able to successfully answer these most basic questions about gravity or some of the other phenomena of physics. Experiments exist that confirm the existence of gravity, but scientists have no underlying theoretical explanation as to why gravity acts as it does.

Gravity did not arise by mutation or natural selection. It was present from the very beginning of the universe. Along with other physical laws, gravity is surely a testimony to a planned creation.

So, then, both religion and science are founded on faith — specifically in the existence of an unseen God (an aspect of religion) vs. the unexplained "hows" and "whys" of many physical laws and, e.g., the apparent ability of material entities to cause non-material aspects of human behavior (an aspect of science).

It's something to think about.

Is Life a Miracle?

When thinking about "the origin of life and humanity," the term "miracle" comes to mind. The dictionary defines "miracle" as a surprising and welcome event that is not explicable by natural or scientific laws and is therefore considered to be the work of a divine agent. A second definition is that a miracle is a highly improbable or extraordinary event, development, or accomplishment that brings very welcome consequences. It would seem that many aspects of the "origin of life" discussed in this book qualify as miracles.

The Bible, the greatest most widely distributed book in the history of mankind, describes many of God's miracles. For example, Moses sure needed God to provide one (i.e., the parting of the Red Sea) when trying to lead the Israelites in escaping from the Egyptian Pharaoh's army.

FIGURE 39: GOD PARTING THE RED SEA FOR MOSES AND THE ISRAELITES

Also, in addition to the miracle of Jesus' conception itself (i.e., inside a virgin), the Bible documents many eye witnessed miracles that Jesus performed. Isn't it then reasonable to believe that God performed miracles including the Big Bang, the beginning of life, the development of flowers and complex organisms, and the creation of consciousness, rational thought, and emotions in humanity?

As Francis Crick, co-discoverer of DNA, Nobel Prize winner, and a philosophical materialist, has conceded: "An honest man, armed with all the knowledge available to us now, could only state that in some sense, the origin of life appears at the moment to be almost a miracle, so many are the conditions which would have had to be satisfied to get it going."[105]

The great physicist and Nobel Lauriat Albert Einstein once said: "There are only two ways to live your life. One is as though nothing is a miracle. The other is as though everything is a miracle."

If Einstein is correct, and given the laws of physics governing the special place that earth occupies in the universe that allows life to exist, the precise tuning of earth's ecosystems, the beauty of flowers, the majesty of mountains, and the very existence of life and humanity (including conscious thought and emotions), the author knows which of Einstein's two options that he is voting for.

Perhaps what is perceived as a miracle regarding the "origin of life" is really yet another extraordinary example of the logical design of the universe by an intelligent Creator of whom there is still much to learn and understand. History suggests so. God has been slowly revealing Himself to mankind for the past few thousand years. Remember the quote from geneticist, physician, Director of the NIH, and Head of the Human Genome Project, Francis Collins, that "DNA is our own instruction book, previously known only to God."[106]

Hopefully there will be more of God's revelations to come. And, just maybe, God is using scientists, including, perhaps ironically, some of the very ones who doubt Him, to accomplish some of this, as they continue learning more about the cell's complex structure and function as well as the sophisticated internal instruction set that governs its operations.

The gap between science and religion may be narrowing, as a new aspect of creation theory has evolved in recent years known as "creation science." This discipline is working on supporting aspects of creation theory via accepted science principles — including mining the constant stream of new data and information coming out of DNA research. For example, by determining the differences in people's DNA and knowing the average chromosomal and mitochondrial DNA mutation rate from one generation to the next, one can determine how far back in time all humans converge to a common ancestor, that being Noah.

Other information now discernible from DNA is where on earth one's ancestors came from. Anyway, some creationists are making use of scientist generated DNA data to help support the case for the creation theory.

In other cases, practices utilized in science (e.g., archeology) have been used in locating and authenticating previously unconfirmed portions of the Bible, further adding to the Bible's credibility. An example is the recent discovery of the correct location and remains of the cities of Sodom and Gomorrah, northeast of the Dead Sea (vs. its previously perceived location to the south). These cities had existed and thrived for many centuries until suddenly, circa 1700 BC, they were destroyed.

Archeologic relics and a thick layer of earthen ash have confirmed the demise of the cities as resulting from a sudden violent fiery (over 2000 deg. F.) sky burst event, consistent with the description of the city's destruction in the Bible's Genesis.[107]

So, perhaps someday, creation theory will be perceived as much a science topic as a religious topic.

Regardless, humans will never totally understand God the Creator. As famous evangelist Billy Graham has stated: "If we try to rationalize God, we will fail. There are mysteries about God that we will never understand. If God can be fully proved by the human mind, then He is no greater than the mind that proves Him" (reference Billy Graham's column from the *Indianapolis Star*, Aug. 17, 2021).

Hopefully this book has encouraged reader interest in the understanding of cell complexity, how the first life might have originated, and how life in humans might have occurred to include conscious thought, emotions, moral behavior, and the ability to create. Such curiosity hopefully prompts thinking about our origin, our purpose on earth, and our destiny.

Public Education Regarding the Origin of Life and Humanity

A s noted in the Introduction, the Creation theory regarding the origin of life is not taught in most public schools due, presumably, to the US Constitution's content on separation of "church and state." Such separation has been supported in certain court decisions.

The apparent basis for the "separation of church and state" is the wording of the Establishment Clause of the Constitution's First Amendment, which "prohibits the government from making any law respecting an establishment of religion. This clause not only forbids the government from establishing an official religion, but also prohibits government actions that unduly favor one religion over another" (ref. US Constitution). A prominent Supreme Court case involving "creation science," known as "McLean v. Arkansas" , was decided in 1982, finding that teaching creationism in public schools violates the First Amendment's Establishment Clause.[108]

Of historical interest is that, until 1968, some states, such as Tennessee, had laws that banned the teaching of evolution. The well-known Scopes trial, which took place in 1925, is often regarded as a landmark in the battle to teach evolution in American public schools. The teacher attempting to teach evolution (based on the recently published work of

Darwin) lost this case, but, eventually, in 1968, the US Supreme Court overturned state laws that banned the teaching of evolution.

A movie was made based on the 1925 Scopes trial titled *Inherit the Wind*. It includes arguments on evolution versus creation, and earned 4 Oscar nominations.

Since 1987, the teaching of creation theory in public schools has been banned (at least as part of the required science curriculum and possibly more broadly), based on a Supreme Court ruling that it is unconstitutional (because it promotes a particular religion). This is problematic and confusing to many people, as the US founding legal document, the Declaration of Independence, acknowledges the existence of a Creator when stating: "…that all men are created equal — that they are endowed by their Creator with certain unalienable Rights, …"

Also, in 1954, the US Congress amended the Pledge of Allegiance (to the American Flag) so that it reads, in part: "…and to the Republic for which it stands, one Nation under God, indivisible, with liberty and justice for all."

Regarding the US Constitution, it deliberately did not include reference to God (or the divine), deferring any such references instead to state constitutions. Subsequently God (or the divine) is mentioned at least once in each of the 50 state constitutions and nearly 200 times overall (Ref. Pew Research Center analysis).

Lyrics to some of our nation's most patriotic songs include references to God. For example, "America the Beautiful" contains the line: "God shed His grace on thee."

The song "God Bless America" speaks for itself.

"The Star-Spangled Banner," our national anthem, contains the line in the 4th stanza: "And this be our motto — In God is our trust."

Also, on July 30, 1956, the 84th US Congress passed a joint resolution declaring "In God We Trust" as the national motto of the United States. The resolution passed both the House and the Senate unanimously. The resolution was passed during the Cold War to, in part, distinguish the US Government from the Soviet Union, which promoted state-sponsored atheism.

Also, in 1956, the 84th Congress passed a new law indicating all currency was to bear the motto "In God We Trust." So, today, all coins and paper currency bear this motto. Apparently, the US Government, as well as all state governments, believe that God exists, but don't trust God enough to allow the teaching of His creations in public schools.

FIGURE 40: ALL MONEY IN THE UNITED STATES (PAPER AND COINS) INCLUDES THE WORDS "IN GOD WE TRUST"

Also, ever since George Washington included ".... so help me God" in his inaugural oath, every president since then has likewise publicly asked for God's assistance at his inauguration.

Of interest also is that prayer is practiced at government sponsored events such as the inauguration of presidents and the opening of legislative sessions. The government even pays to have chaplains on its staffs.

So, it appears a contradiction exists in that the "creation" theory cannot be taught in public schools due to separation of church and state, even though the US government (i.e., the national state), via various official documents, the National Motto, a notation on all currency, and some of their own ceremonies and prayer practices, acknowledges the existence of God, the Creator and/or in God whom we trust. Therefore, many of government's own practices and legislative products do not suggest a separation of church and state.

But some of government's pronouncements do suggest such a separation. For example, the government has outlawed teacher led prayers, Bible readings, posting of the Ten Commandments, and mandatory "moments of silence" in public schools. They have also outlawed religious displays (e.g., Nativity scenes) on public property.

Much confusion has subsequently resulted on the rationale of what is allowed and what is not, with court cases common place. That is, a definitive description does not seem to exist that clearly defines the "line in the sand" regarding separation of church and state?

As an example of the issues in the courts and as a ray of hope regarding judicial decisions applicable to schools that offer religious teaching, the US Supreme Court, on June 21, 2022 declared unconstitutional a Maine state law that prohibited using public money for schools that offer religious instruction. The majority opinion (for the 6-3 vote) noted that the Maine law represented discrimination against religion, which violates the First Amendment.

Regarding public education with respect to the origin of life, knowing that science (or the statements of science-based groups) is not always correct, and when major differences of opinion exist on a topic, shouldn't all major positions on a controversial topic be presented in an educational environment, noting the pros and cons of each — at least until one position is proven to be the correct one?

The presentation of pros and cons should include any assumptions, data and discussion on the validity of assumptions, and perhaps estimated probabilities of a particular position being correct. For example, the section in this book dealing with statistical probabilities notes that most published estimates of natural phenomena being the cause of the origin of life indicate the probability of its being true are essentially zero. So, why is natural phenomena being the cause of the origin of life being taught in schools as factual?

In a courtroom or classroom oral debate — or on the editorial page of a newspaper, both sides of a controversial topic are presented. That's not how the origin of life is being taught. Science textbooks generally just present a theoretical description of natural phenomena as the cause of the origin of life; there is little or no content presenting the many reasons to question this hypothesis, including its likelihood of it being correct as essentially zero.

This book, in part, serves to mention several of the reasons to question the natural phenomena theory. Could it be that both theories have elements of truth, with creation responsible for the origin of life and some combination of creation and evolution taking over from there?

One purpose of the educational system is to develop in students the ability to think logically with an inquisitive inquiring mind, using data when available and as appropriate. Wikipedia (the pseudo-encyclopedia of the Internet) defines education as "the process of acquiring knowledge, skills, values, beliefs, and habits." Presenting only one side of a controversial topic, when little, if any, confirming data exists for its validity, would seem to conflict with some of these educational objectives.

Further, when science is taught in schools, utilization of the official standard "scientific method" is always emphasized, i.e., to propose a hypothesis and then design experiments to prove or disprove it. That is not how the origin of life topic is being presented in public education. It's as if scientists in academia have already decided what they think the right answer is — and are presenting it as truth, despite the lack of supporting data, confirming experiments, and reasonable probability of it being true. In fact, the creation theory appears to be the better fit to the evidence that humans have about their origin, intelligence, and humanity.

If presenting only one theory wasn't concern enough, public education is also still often presenting that theory as to how it was understood in past decades (when living cells were thought to be simple structures). Origin of Life experiments and philosophy have evolved significantly in recent years. The book *The Mystery of Life's Origin, the Continuing Controversy* brings the discussion up to 2020 and has an entire chapter titled "Textbooks Still Misrepresent The Origin of Life." It's an interesting update on the state of the controversy.[109]

So, as of 2024, natural phenomena causing the origin of life is still just a theory. If the creation theory is not included in the science classroom, so be it. But the natural phenomena theory should not be presented as factual (since it is not even close to having been established as truth). Also, the reasons for questioning the natural phenomena theory should be included in its teaching, so as to promote among students an open unbiased view of the theory, representative of the current state of the debate.

So, regarding public education, as long as natural phenomena is considered a possible option by leaders of the scientific community to explain the origin of life, and even though it is unproven and statistically unlikely, it is what is taught in most public schools as fact. And most students tend to believe (and not challenge) what they are taught.

In pondering the rationale for this perceived one-sided presentation of the origin of life by the science community, natural phenomena and evolution, which are science topics, are taught as part of the required science portion of a school's curriculum. Creationism, while arguably a better fit to the evidence, is, at least partly, a religion topic and religion is not a required course in public schools. In fact, it is not even an elective course in most public schools. Also, most adults, including most creationists, accept evolution as explaining many of the changes that have occurred in organisms over time (at least within a species of organisms).

Some of what remains controversial regarding the natural phenomena theory is whether evolution can explain and prove how one species of organisms can evolve to another (e.g., chimpanzees to humans), since the existence of such evolution would result in a large number of transitional organisms in both habitat space and time. Such transitional forms of organisms don't seem to exist either in fossils or present life.

Darwin himself in his *On the Origin of Species* (1859), titled Chapter 6 "Difficulties of the Theory". He states: "Firstly, why, if species have descended from other species by insensible fine gradations, do we not everywhere see innumerable transitional forms? Why is not all nature in confusion instead of the species, as we see them, being well defined? "

Said another way, different species have different information contained in their DNA. Mutations (with DNA mutations being nature's way of making changes) are not known as a way of adding information to DNA (as would be required, e.g., in evolving from one species to a higher order species such as from apes to humans). Biologists might want to address this topic in the next edition of their textbooks.

A review was done of many biology textbooks a few years ago.[110, 111] One conclusion was: "The majority of textbooks exude confidence that confirmation of a naturalistic model of life's origin is inevitable." Also cited in the literature is the conclusion: "More than fifty years of experimentation on the origin of life in the fields of chemical and molecular evolution have led to a better perception of the immensity of the problem in explaining the origin of life on Earth — rather than to its solution. At present, all discussions on principle theories and experiments in the "origin of life" field either end in stalemate or in a confession of ignorance."[112]

As a specific example, most textbooks still present the famous Miller experiment as though it reflects the earth's early environment (i.e., reducing in nature, thus allowing for chemical reactions resulting in amino acids), whereas most geochemists since the 1960s suggest that earth's early environment was totally unlike that assumed and used by Miller.[113]

Which brings us back to one of the additional issues prompted by this book: "Why does the public education system insist on presenting only one theory regarding the origin of life, with that theory presented as truth, when it is unlikely, unproven, highly speculative, seemingly inconsistent with the most recent research, and when at least one other theory is popularly accepted by much of mankind?

One definition of a scientific theory is that it is a broad explanation of a natural phenomenon that is testable and based on extensive experimentation. So where is the test data providing convincing evidence in support of the "natural phenomena" theory? And, as noted earlier, one of the greatest scientific leaders of all time, Louis Pasteur, Father of Biogenesis, has provided scientific evidence in concluding that "life can only come from other life and not from non-living materials" which refutes the theory, taught in public schools, that life began from non-life chemicals.

Famous Scientists
Who Were Also Devout Christians

W hile the stereotyped scientist and educator is perceived as believing the natural phenomena theory of the origin of life (or of having no opinion), PEW surveys have suggested the scientific community is split roughly 50: 50 as to believing in a God (or some higher power than humans).

Below is a partial list of well-known scientists that are also known to have been devout Christians:[114]

Isaac Newton — discovered gravity and invented calculus

Johannes Kepler — astronomer and mathematician, laws of planetary motion

Robert Boyle — chemist, ideal gas law

James Maxwell — electromagnetism

Blaise Pascal — mathematician, theory of probabilities, calculators

Wilhelm Leibniz — mathematician, calculus

Carl Linnaeus — botanist, nomenclature, father of taxonomy

Antonie van Leeuwenhoek — microscope, father of microbiology, bacteria

Joseph Priestley — chemist, discovery of oxygen

Charles Babbage— computer, father of computing

Gregor Mendel — genetics

Michael Faraday — electric motor, electric generator

Samuel Morse — telegraph, Morse Code

Alessandro Volta — physicist, electricity, battery

Andre-Marie Ampere — physicist, electromagnetism, solenoid

Heinrich Hertz — physicist, radar, electrical frequency

James Joule — physicist, energy, 1st law of thermodynamics

George Washington Carver — inventor, crop rotation

Igor Sikorsky — helicopters, aircraft design

Wernher von Braun — space rockets, father of rocket science

Werner Heisenberg — quantum physics

Francis Collins — Human Genome Project, National Institutes of Health Director

Of course, the list would be much longer if we were to include famous Jewish scientists.

Glossary of Biology Terms

Abiogenesis: the idea that life arose from non-life more than 3.5 billion years ago on Earth (subsequently thought to be disproved by the work of Louis Pasteur).

Agnostics: people who believe that the existence of God, of the divine or the supernatural, is unknown or unknowable.

Amino Acid: one of about 20 different molecules used as the building blocks of proteins

Athiests: people who don't believe in the existence of deities (e.g., God).

ATP: (Adenosine triphosphate) is an organic compound that stores and provides energy to drive many chemical processes in living cells

Biogenesis: the hypothesis that living matter arises only from other living matter.

BioMolecule: a molecule present in organisms that is essential to one or more biological processes.

Black Hole: a place in space where gravity is so strong (due to matter being squeezed into a tiny space) that even light cannot get out. This can happen when a star is dying (i.e., collapsing). Because no light can get out, black holes are invisible. They are detected by the deflection of light from other sources passing nearby (and also by measuring the orbits of nearby celestial bodies).

Catalyst: a substance that enables a chemical reaction to proceed at a different rate (usually faster) or under different conditions (e.g., lower temperature) than otherwise occurs.

CRISPR: an acronym for "clustered regularly interspaced short palindromic repeats" which represents a family of DNA sequences found in the genomes of prokaryotic organisms such as bacteria. CRISPR is a way cells use to find a specific bit of DNA inside a cell.

DNA: (deoxyribonucleic acid) is the molecule, consisting mostly of nucleic acids, that contains the genetic code of organisms. DNA is in each cell in an organism and instructs cells, e.g., on how to make specific proteins.

Eukaryotic Cells: cells typical of plants and animals, — more complex than smaller prokaryotic cells such as bacteria. Such cells have a defined nucleus and contain more organelles (e.g., mitochrondria, lysosomes) than prokaryotic cells.

Evolution: the change in the inheritable characteristics of biological populations over successive generations.

Gene: a sequence of nucleic acids in DNA that codes for a specific protein. A gene is the basic physical and functional unit of heredity.

Homochirality: use of a single isomer type in constructing a molecule. Objects are chiral when they cannot be superimposed on their mirror images. Proteins and nucleic acids are homochiral (i.e., have isomer purity) since they only contain left (for proteins) or right (for nucleic acid) handed building block isomers.

Hydrolysis: the chemical breakdown of substances by water, depending on conditions such as chemistry, solubility, pH, and oxidation-reduction potential.

Isomer: molecules with identical molecular formulas (i.e., the same number of atoms of each element) but distinct arrangements of atoms in 3-dimensional space.

Lipid: a type of organic molecule found in living things, composed of fats and oils. It is a main component of cell membranes and cell walls.

Metabolic Map: a chart showing the different chemical reactions that occur inside a cell.

Metabolism: the set of life-sustaining chemical reactions in cells. The purposes of metabolism are: the conversion of food to energy to run cellular processes; the conversion of food/fuel to building blocks for proteins, lipids, nucleic acids, and some carbohydrates; and the elimination of nitrogenous wastes.

Mitochondria: membrane-bound modules within cells that generate most of the chemical energy needed to power the cell's biochemical reactions. Chemical energy produced by the mitochondria is stored in a small molecule called adenosine triphosphate (ATP).

Nucleic Acid: a molecule (e.g., DNA and RNA) consisting of a sequence of nucleotides.

Nucleotide: one of 5 different molecules used as the building blocks of nucleic acids RNA and DNA. *Note: adenine, cytosine, and guanine are building blocks in both RNA and DNA, while thymine occurs only in DNA and uracil only in RNA.*

Organic Molecule: a carbon-based compound. Organic compounds contain carbon bonds in which at least one carbon atom is linked to an atom of another type (usually hydrogen, oxygen or nitrogen). Most molecules in cells are organic molecules.

Oxidizing Environment: an environment (e.g., atmosphere) in which atoms and molecules bond with oxygen atoms and, therefore, lose one or more electrons. When a compound loses an electron, it is oxidized. So, $Fe + O_2 -> FeO_2$ is an oxidizing reaction as Fe gives up 2 electrons in the reaction. An oxidizing atmosphere makes producing organic compounds impossible.

Peptide: a chain of 2-50 amino acids (i.e., smaller than a protein). It normally has a simple function (vs. more complex functions for proteins). It functions usually by virtue of its primary structure while proteins usually require secondary, tertiary, and sometimes quaternary structure to function properly.

Plasmids: independent, circular, self-replicating DNA molecules that carry only a few genes, none of which are critical to the cell's ability to grow, sustain life, and reproduce. Plasmids are typically found in single cell organisms without a well-defined nucleus (i.e., prokaryotes). They are separate from the main source of DNA, that being the single chromosome DNA located in the nucleoid.

Precursor: a substance from which another substance is formed. E.g., a chemical compound preceding another in a metabolic pathway. E.g., in the manufacture of insulin, the precursor is proinsulin, which is the form of the molecule that goes through a conversion to the final form of insulin.

Prokarytoic Cells: small unicellular organisms (e.g., bacteria) thought to be representative of the first life forms on earth. While they contain DNA, they do not have a well-defined nucleus, nor do they have some of the membrane-bound organelles characteristic of animal and plant cells.

Protein: a large biomolecule consisting of one or more long chains of amino acid "building blocks" (typically totaling 50 or more amino acids) whose function depends on the specific sequence of amino acids and on the molecule's folding structure.

Redox Chemical Reactions (i.e., oxidation reduction reactions): The oxidation of a molecule involves the removal of an electron from its structure, while reduction is the addition of an electron to a molecule. When electrons are removed from a molecule, they are transferred to another molecule, since they do not exist as free entities. Therefore, when one molecule is oxidized another must simultaneously be reduced, hence the term redox to describe oxidation/reduction reactions. The chemistry of life is mostly about the chemistry of redox reactions of carbon compounds. The significance of redox reactions lies in the energy transformations that take place during the course of the reactions. It is the free energy liberated by redox reactions that is harnessed by organisms to build, maintain, and reproduce themselves.

Reducing Atmosphere/Environment: an atmospheric condition/environment in which oxidation is prevented by lack of or removal of oxygen and other oxidizing gases and which may contain actively reducing gases such as hydrogen, carbon monoxide, and hydrogen sulfice that would be oxidized by any present oxygen. A reducing atmosphere is one in which atoms and molecules bond with hydrogen rather than oxygen.

Ribosome: the module within a cell where assembly of proteins takes place.

RNA: (Ribonucleic acid) — a large biomolecule, consisting primarily of nucleic acids, essential in various biological roles in coding, decoding, regulation and expression of genes.

Singularity: an irregularity in a mathematical model or function; e.g., a point at which a continuous function becomes discontinuous. Often, a singularity is a point in which a mathematical equation's computational result is infinity. Also, a singularity is a point at which a given mathematical object is not defined.

Spontaneous Generation: the supposed production of living organisms from nonliving matter.

Synthesis: the process of two or more chemical species combining to form a more complex product: e.g., A + B -> AB

TCA Cycle (Tricarboxylic Acid Cycle, also known as the Citric Acid Cycle or Krebs Cycle): the series of chemical reactions occurring within cells of all aerobic organisms (i.e., those consuming oxygen) to store and release energy. The TCA cycle is a standard part of the metabolism of cells.

Acknowledgments

The author thanks the following individuals for their review of this book:

Ms. Martha Alford (wife of the author, registered dietitian) Dr.

Richard Baltz, (molecular biologist)

Rev. James Gentry, (pastor)

Dr. Matthew Hilton (industrial microbiologist)

Rev. Glenn McDonald (pastor, former molecular biologist) Dr.

William Muth (industrial microbiologist)

Mrs. Laura Rain (ordained spiritual counselor)

Dr. James Riggs (Professor of Chemical Engineering)

Dr. Quentin Small (pastor)

Mr. Joel Robichaud, (chemistry instructor, Champlain College)

Dr. Andy Russell, (bioengineer)

Ms. Gail South, (editorial staff for various publications)

Dr. Leif Solberg, M. D.

Mr. Tom Tower (Air Force, Lt. Colonel, lay minister)

Dr. Joanne Towery, (food scientist and Christianity advocate)

Rev. David Williamson (pastor)

Dr. Greg Wise, M. D.

Endnotes

1   Bird, W. R. *The Origin of Species Revisited,* Thomas Nelson Co., Nashville, 1991, pp. 298-99

2   Harold, Franklin M., *The Way of the Cell: Molecules, Organisms, and the Order of Life,* Oxford University Press; First Edition (June 5, 2003)

3   Metaxas, Eric, *Is Atheism Dead,* Salem Books (October 19, 2021)

4   Ross Hugh *The Creator and the Cosmos: How the Latest Scientific Discoveries Reveal God* RTB Press; 4th edition (March 1, 2018)

5   Ibid.

6   Opinion Differences between Public and AAAS Scientists, Pew Research Center survey, 2014.

7   "Scientists and Beliefs, Religion & Science in the United States," Pew Research Center report, Nov. 5, 2009

8   Stenger, Victor J. *God: The Failed Hypothesis. How Science Shows That God Does Not Exist,* Promethius, 1st Edition, January 2017

9   Andrews, Egar H. *Who Made God?* 3rd edition, Evangelical Press (March 2, 2016)

10   Oppenheimer, Robert J. and Snyder Hartland, "On Continued Gravitational Contraction," American Physics Society, 1939

11   Wikipedia explanation of The Big Bang.

12   Hawking, Stephen *A Brief History of Time,* Random House Publishing Group; 10th Anniversary edition (September 1, 1998) Sept. 1, 1988

13   Ross, Hugh *The Creator and the Cosmos, RTB Press, 2018*

14 Strobel, Lee *The Case for a Creator: A Journalist Investigates Scientific Evidence That Points Toward God* Zondervan; Reprint edition (February 25, 2014)

15 Ross, H. *The Creator and the Cosmos*

16 Ibid.

17 Ham, Ken and Hodge, Bodie *The New Answers: Creation/Evolution and the Bible*, Book 2, Master Books; Illustrated edition (May 14, 2008)

18 Rees, Martin *Just Six Numbers: The Deep Forces That Shape The Universe* Basic Books; First American Edition (May 3, 2001)

19 Ross, H. *The Creator and the Cosmos*

20 Strobel, L. *The Case for a Creator*

21 Ibid.

22 https: //www. allaboutscience. org/scientists-and-the-cosmological-arguments-faq. htm#: ~: text=Wernher%20von%20Braun%20 (Pioneer%20rocket, deny%20the%20advances%20of%20science.%22

23 Andrews, Edgar *Who Made God?* Evangelical Press (May 24, 2012)

24 The journal *Science,* 1953

25 Wells, Jonathan; Thaxton, Charles; Bradley, Walter; Olsen, Roger; Tour, James; Meyer, Stephen; Gonzalez, Guillermo; Miller, Brian; Klinghoffer, David *The Mystery of Life's Origin, the Continuing Controversy* Discovery Institute (January 27, 2020)

    (Note: This reference includes a summary of the major experiments and revised thinking among scientists regarding the Origin of Life that has occurred in recent years (i.e., up to 2020).

26 Trail, Dustin; Watson, E. Bruce; Tailby, Nicholas D. *The oxidation state of Hadean magmas and implications for early Earth's atmosphere.* Nature 480, 79–82 (2011). https://www.nature.com/articles/nature10655

27 Ham and Hodge *The New Answers Book: Over 25 Questions on Creation/ Evolution and the Bible,* Book 1, Master Books; Illustrated edition (December 10, 2006)

28 Wells, Thaxton, Bradley, Olsen, Tour, Meyer, Gonzalez, Miller and Klinghoffer, *The Mystery of Life's Origin, the Continuing Controversy*

29 Ibid.

30 Ibid.

31   Ibid.

32   Hazen, Robert M. *The Origins of Life [The Great Courses, 2005] Course Guidebook, www.thegreatcourses.com*

33   Denton, Michael *Evolution: A Theory In Crisis* Adler & Adler Pub; 3rd edition (April 15, 1986)

34   Andrews, E. *Who Made God?*

35   Ibid.

36   Hazen, R. *Origins of Life*

37   Strobel, L. *The Case for a Creator*

38   Hazen, R. Origins of Life, The Great Courses series

39   Andrews, E. *Who Made God?*

40   Ibid.

41   Haldane, J. B. S. "Data Needed for a Blueprint of the First Organism," *Journal of Genetics,* Vol. 96, No. 5, published online November 2017, pp. 735–739 Indian Academy of Sciences https: //doi. org/10. 1007/s12041-017-0831-6

42   Isaacson, Walter *The Code Breaker: Jennifer Doudna, Gene Editing, and the Future of the Human Race,* Simon and Schuster, 2021

43   https: //www. snexplores. org./article/explainer-how-crispr-works

44   Hazen, R. *Origins of Life*

45   Ross, H. *The Creator and the Cosmos*

46   Hoyle, Fred *Intelligent Universe: A New View of Creation and Evolution,* Holt Rinehart & Winston; First Edition (January 1, 1983)

47   Andrews, E. *Who Made God?*

48   First Things, Oct. 2000 (https //www. firstthings. com/article/2000/10/ how-intelligent-is-intelligent-design)

49   Shapiro, Robert "A Simpler Origin for Life," *Scientific American,* Feb. 12, 2007.

50   Hazen, R. *Origins of Life*

51   Strobel, L. *The Case for a Creator*

52   Ham and Hodge, *The New Answers, Creation/Evolution and the Bible*

53   Strobel L. quoting Michael Behe, Ph.D., in *The Case for a Creator*

54 Ham and Hodge, *The New Answers, Creation/Evolution and the Bible*

55 www.ideacenter.org: Michael Behe, Irreducible Complexity: The Challenge to the Darwinian Evolutionary Explanations of Many Biochemical Structures. 2004.

56 Strobel L., quoting Michael Behe, Ph.D., in *The Case for a Creator*

57 Yockey, Hubert P. "Self Organization Origin of Life Scenarios and Information Theory," *Journal of Theoretical Biology* Volume 91, Issue 1, 7 July 1981, Pages 13-31

58 Johnston, George Sim *Did Darwin Get It Right? : Catholics and the Theory of Evolution* Our Sunday Visitor; First Edition (January 1, 1998)

59 Yockey, H, P. *Self Organization Origin of Life Scenarios and Information Theory*

60 Stenger, V. *God, The Failed Hypothesis*

61 Andrews, E. *Who Made God?*

62 Collins, Francis S. *The Language of God: A Scientist Presents Evidence for Belief,* Free Press, 2007. (Collins discovered the genes associated with a number of diseases and led the Human Genome Project. He served as director of the National Institutes of Health (NIH) in Bethesda, Maryland, from 17 August 2009 to 19 December 2021. Before being appointed director of the NIH, Collins led the Human Genome Project and other genomics research initiatives as director of the National Human Genome Research Institute.)

63 Andrews, E. *Who Made God?*

64 Ham and Hodge, *The New Answers, Creation/Evolution and the Bible*

65 Ibid.

66 Lester, Lane P. and Bohlin, Raymond G. T*he Natural Limits to Biological Change* Probe Ministries Intl; Subsequent edition (January 1, 1989)

67 Andrews, E. *Who Made God?*

68 Strobel, L. *The Case for a Creator*

69 Ibid.

70 Ibid.

71 Ibid

72 Collins, F., *The Language of God: A Scientist Presents Evidence for Belief,* Free Press, 2007.

73   Strobel, L. *The Case for a Creator*

74   Meyer, Stephen C. (2001), *Word Games: DNA, Design and Intelligence, in Signs of Intelligence*, W. A. Dembski and James Kushiner, edts. (Brazos Press: Grand Rapids, MI.) pp. 102-111

75   Stenger, V. *God: The Failed Hypothesis*

76   https: //embryo. asu. edu/pages/hayflick-limit

77   Wells, Thaxton, Bradley, Olsen, Tour, Meyer, Gonzalez, Miller and Klinghoffer, *The Mystery of Life's Origin, the Continuing Controversy*

78   Ham, K. *The New Answers, Creation/Evolution and the Bible*

79   Borel, Emil *Probabilities and Life*, Dover Publications, 1962

80   Ham and Hodge *The New Answers, Creation/Evolution and the Bible*

81   Hoyle, Sir Fred and Wickramasinghe, Chandra *Evolution from Space: A Theory of Cosmic Creationism*, Simon & Schuster; 1981

82   Sagan, Carl (editor), *Communication with Extra-Terrestrial Intelligence*, MIT Press, 1973

83   Cedar Creek Community Church publication, "An 'Expert' Opinion," p. 13, quotes Richard Peacock's article titled "The Probability of Life" that appeared on his now-defunct website, evolutionfaq.com

84   Strobel, L. *The Case for a Creator*

85   Salisbury, Frank B. "Doubts About the Modern Synthetic Theory of Evolution," *American Biology Teacher*, 33, 6, pp. 335-338, 1971.

86   Morris, Ph.D., Henry M. "The Mathematical Impossibility of Evolution," 2003, Institute for Creation Research No. 179, icr. org

87   Andrews, E. *Who Made God?*

88   Ham and Hodge *The New Answers, Creation/Evolution and the Bible*

89   Prigogine, Llya; Nicolis, Gregoire; Babloyantz, Agnes "Thermodynamics of Evolution" *Physics Today* 25, (12), 38–44 (1972) ; https: //doi. org/10. 1063/1. 3071140

90   Hoyle, F. and Wickramasinghe, N. C. *Evolution from Space*

91   Ross, H. *The Creator and the Cosmos*

92   Andrews, E. *Who Made God?*

93 L. Strobel, *The Case for a Creator*

94 Wilder Penfield, *The Mystery of the Mind: A Critical Study of Consciousness and the Human Brain,* Princeton Legacy Library, 1979

95 Strobel, L. *The Case for a Creator*

96 Ross, H. *The Creator and the Cosmos*

97 Ross, Hugh *Creation and Time: A Biblical and Scientific Perspective on the Creation-Date Controversy,* Navpress Pub Group, 1994

98 Ham and Hodge, *The New Answers, Creation/Evolution and the Bible*

99 *Unlocking the Mystery of Life,* Dean Kenyon's section of the 67-minute 2003 documentary

100 *Science Alert,* March 2015

101 Wells, Thaxton, Bradley, Olsen, Tour, Meyer, Gonzalez, Miller and Klinghoffer, *The Mystery of Life's Origin, the Continuing Controversy*

102 Andrews, E. *Who Made God?*

103 www. worldreligions. info

104 Davies, Paul "Taking Science on Faith," Opinion, *New York Times,* Nov 24, 2007

105 Crick, Francis *Life Itself: Its Origins and Nature* Simon and Schuster 1981

106 Collins, F., *The Language of God: A Scientist Presents Evidence for Belief,* Free Press, 2007.

107 Metaxas, E. *Is Atheism Dead?*

108 Wells, Thaxton, Bradley, Olsen, Tour, Meyer, Gonzalez, Miller and Klinghoffer, *The Mystery of Life's Origin, the Continuing Controversy*

109 Ibid.

110 Mills, Professor Gordon C.; Lancaster, Professor Malcolm; Bradley, Professor Walter L. "Origin of Life and Evolution in Biology Textbooks: A Critique," *American Biology Teacher,* Volume 55, No. 2, February 1993, 78-83

111 Ibid.

112 Strobel, L. *The Case for a Creator*

113 Ibid.

114 Metaxas, E. *Is Atheism Dead?*

About the Author

Joseph Alford, a former Navy Officer, has BS, MS, and PhD degrees in Chemical Engineering and spent most of his career working alongside experts in genetic engineering and microbiology while helping develop and automate life science processes (i.e., utilizing microorganisms) for manufacturing pharmaceutical products (e.g., antibiotics, insulin) for the benefit of society. He is an elected Fellow of both the American Institute of Chemical Engineers and the International Society of Automation.

He has been elected as a Distinguished Alumni of the College of Engineering of both his alma maters, is a member of the Process Automation Hall of Fame, and has served as a Deacon and Elder in the Presbyterian Church. In semi-retirement, he continues consulting, giving guest lectures at universities, and serving on several Advisory Boards.

Correspondence with the author can be directed to:
jmalford5@earthlink. net